KB039996

THE ESSENTIAL GUIDE TO
ALGEBRA 1

유하림(Harim Yoo) 지음

HERMONHOUSE

Preface

To. 학부모님과 학생들께

압구정 현장 강의를 통해 Prealgebra부터 AP Calculus까지 가르치면서, 강사로서 느낀 점은, 학부모님들이 선택할 수 있는 국내에서 판매되는 전문 교재가 몇 개 없다는 것과 충분히 많은 연구와 노력을 통해 개발된 교재들은 더더욱 부족하다는 점이었습니다.

국내 수능 시장에서는 "OOO" 커리큘럼으로 진행되는 수업들과 교재들이 많은 데 비해, 미국 SAT 시장과 유학 시장은 그렇지 않았기 때문에, 이렇게 유하림 커리큘럼의 시작을 알리는 것이 기대되면서도 떨립니다.

Algebra 1을 처음 배우는 학생들이 흥미롭게 배우길 희망하면서 교재를 썼습니다. 또한, 강의나 교재가 너무 쉽다고 느껴지진 않을 정도의 개념 강의 수준을 유지하기 위해 노력했으며, 지금까지 수업해왔던 내용들을 많이 담았기 때문에 국내에서 유학을 준비하는 학생이나 해외에서 공부하고 있는 학생들에게 도움이 되길 진심으로 희망해봅니다.

이 교재를 출간할 수 있도록 물심양면으로 힘써주신 마스터프랩 권주근 대표님께 감사합니다. 또한 저에게 항상 롤모델이 되어주고, 강사로서 성장할 수 있는 원동력을 주시고 계신 심현성 대표님께도 감사의 마음을 전합니다. 그리고 언제나 든든한 지원군인 제 아내와 딸, 부모님께도 항상 감사합니다. 마지막으로, 제 삶에 이러한 기회를 주신 하나님께 감사드립니다. 앞으로도 더 좋은 교재를 만들어 견고하고 튼튼한 유하림 커리큘럼을 완성시키길 희망합니다.

2020년 봄
유하림

저자 소개

유하림(Harim Yoo)

미국 Northwestern University,
B.A. in Mathematics and Economics
(노스웨스턴 대학교 수학과/경제학과 졸업)

마스터프렙 수학영역 대표강사
압구정 현장강의 ReachPrep 원장

고등학교 시절 문과였다가, 미국 노스웨스턴 대학교 학부 시절 재학 중 수학에 매료되어, Calculus 및 Multivariable Calculus 조교 활동 및 수학 강의 활동을 해온 문/이과를 아우르는 독특한 이력을 가진 강사이다. 현재 압구정 미국수학/과학전문학원으로 ReachPrep(리치프렙)을 운영 중이며, 미국 명문 보딩스쿨 학생들과 국내 외국인학교 및 국제학교 학생들을 꾸준히 지도하면서 명성을 쌓아가고 있다.

2010년 자기주도학습서인 "몰입공부"를 집필한 이후, 미국 중고교수학에 관심을 본격적으로 가지게 되었고, 현재 유하림커리큘럼 Essential Math Series를 집필하여, 압구정 현장강의 미국수학프리패스를 통해, 압도적으로 많은 학생들의 피드백을 통해, 발전적으로 교재 집필에 힘쓰고 있다.

유학분야 인터넷 강의 1위 사이트인 마스터프렙 수학영역 대표강사 중 한 명으로 미국 수학 커리큘럼의 기초수학부터 경시수학까지 모두 영어와 한국어로 강의하면서, 실전 경험을 쌓아 그 전문성을 확고히 하고 있다.

[저 서]   몰입공부
         The Essential Workbook for SAT Math Level 2
         Essential Math Series 시리즈

저자직강 인터넷 강의는 SAT, AP No.1 인터넷 강의 사이트인 마스터프랩 (www.masterprep.net) 에서 보실 수 있습니다.

이 책의 특징

유하림 커리큘럼 Essential Math Series의 두번째 책입니다. The Essential Guide to Prealgebra 교재를 통해 공부를 마친 학생들을 위해 집필한 교재입니다. 미국 명문 Junior Boarding School 및 Boarding School을 진학하고, 성공적으로 적응하기 위해 반드시 필요한 내용이 무엇일까 고민하였고, 7학년(예비8학년)에게 가장 필요한 Algebra 1 교재가 무엇일까 생각하면서 고민하고 수정하며 작업한 교재입니다.

1st

기본에 충실한 책

이 책을 통해, 혼자서 고민하고 공부할 학생들과 마스터프렙 인강을 통해 공부할 학생들, 그리고 현장강의를 통해 저와 함께 공부할 학생들을 위해, 기본에 가장 충실한 교재를 집필하고자 노력했습니다. Prealgebra 이후 수학과목을 듣는 학생들을 위해, 이 교재를 통해 Algebra 1 과정에서 반드시 필요한 예제들을 포함시켜 집필하고 개정하는 과정을 여러 번 반복했습니다. 학생의 눈높이에 맞추기 위해, 현장강의 및 인터넷 강의에 대한 학생들의 다양한 반응을 토대로 지속적으로 수정한 노력의 산물입니다.

2nd
생각의 확장을 위한 책

Essential Math Series를 AMC와 같은 경시수학을 준비하려는 학생을 위한 시리즈로 만들기 위해 예제 선정에 고민하고, 풀이방향을 잡았습니다. 교과 과정의 기본과 더불어 문제 해결의 가장 본질이 되는 개념을 어떻게 설명하고, 어떻게 받아들여야 심화 문제에서 생각을 확장할 수 있을지 고민하며 집필하였습니다. 특히, Algebra 1의 경우, Word Problems과 Quadratic Function과 Equation을 바라보는 올바른 관점과 생각의 필수재료를 최대한 담기 위해 노력했습니다.

3rd
유학생을 위한 단 한 권의 책

미국수학을 정말 미국수학답게 가르치기 위해 열심히 공부하고 연구하고, 앞으로도 그러할 것입니다. 노스웨스턴 대학교 학창시절 수학에 대한 열정을 뒤늦게 꽃피워 밤새워 공부했던 것처럼, 저는 학생들을 더 잘 가르치고, 더 나은 미래로 이끌기 위해, AMC, AIME, ARML, HMMT, PUMaC, SUMO와 같은 문제들을 동일한 열정으로 밤낮없이 풀고 해석합니다. 여러분이 지금 보는 이 책은 제 현재 노력의 최선의 산실이며, 앞으로도 그러할 것입니다. 이 책을 통해 수학을 두려워하지 않고, 문제 해결을 즐거워하며, 이른 나이에 수학에 대한 열정을 꽃피우길 기대합니다.

CONTENTS

Topic 1

Mathematical Expressions

1.1 Mathematical Expressions

A variable, usually denoted by x, is a placeholder. It is distinct from a constant, which only keeps one value. When we combine numbers and/or variables using operations such as addition, multiplication, division, and subtraction, we form a mathematical expression.

Because x is such a commonly used variable, we will no longer use '$\times$' for multiplication. Instead, we will sometimes denote products using '$\cdot$', so that $3 \cdot 4$ equals 3 times 4. Sometimes, the product can be indicated by simply putting two expressions in parentheses next to each other, or just putting a constant(=number) next to an expression in parentheses:

$$(4+5)(6-3) \quad \text{means} \quad (4+5) \cdot (6-3),$$
$$4(6-19) \quad \text{means} \quad 4 \cdot (6-19).$$

When a constant is multiplied by a variable, we say that the constant is the coefficient of the variable, so that 6 is the coefficient of x in the expression $6x$. We call the product of a constant and a variable raised to some power a term. A constant by itself is a term as well, so in the expression $3x+7$, both $3x$ and 7 are considered terms.

Example

Evaluate the expression $x^2 - 3x + 2$ when $x = 1$.

Solution
Evaluation means *substitution*. If we substitute $x = 1$ into the original expression, then $1^2 - 3(1) + 2 = 1 - 3 + 2 = 0$. Hence, the answer is 0.

$\boxed{1}$ Evaluate each of the following when $x = 6$.

(a) $(x+12)/x$ (b) $3x^2$ (c) $\sqrt{5x-5}$

1.2 Arithmetic with Expressions

First, when we perform arithmetic with mathematical expressions, we *group* similar terms together. For instance, given an expression $3x + 4 + 5x - 3$, $(3x + 5x) + (4 - 3) = 8x + 1$ is the way we group terms together.

Example

Simplify the following expressions.

(a) $3a + 2 - 4a - 7$ (b) $2x + 3y + 4x - 8y$

Solution

(a) Simplification means *grouping* similar terms. Hence,

$$3a + 2 - 4a - 7 = (3a - 4a) + (2 - 7)$$
$$= -a - 5$$

(b) If there are two variables, unless specified, consider them all distinct. Hence,

$$2x + 3y + 4x - 8y = (2x + 4x) + (3y - 8y)$$
$$= 6x - 5y$$

2 Simplify each of the following expressions.

(a) $(4x - 3) + (6x - 7)$ (b) $(2 - 5x) + (-17x - 27)$

Second, when we see exponential expression or power expression, remember that the exponent tells us the number of times we multiply the base. For instance,

$$\underbrace{a \times a \times a \times a \times a}_{5 \text{ times}} = a^5$$

Example

Write the following expression in the simplest form.

$$a^2 a^{-3} a^5$$

Solution

The positive exponent tells us the number of times the base appears. The negative exponent, on the other hand, must be changed into the reciprocal to use the property of positive exponent. Hence,

$$a^2 a^{-3} a^5 = a^2 \times \frac{1}{a^3} \times a^5 = \frac{a^7}{a^3} = a^4$$

3 Simplify each of the following expressions.

(a) $a^3 \cdot a^4$ (b) $(8x^3)(18x^2)$ (c) $(2x^3)^4$

1.3   Manipulation of Expressions

If we reverse the process of distributive property, we call it factoring. Let's have a look at what distributive property is.

$$\bigcirc \times (\square \pm \triangle) = \bigcirc \cdot \square \pm \bigcirc \cdot \triangle$$

4 Factor $-25a^3 + 40$.

5 Factor the expression $x(3x+2) + 2(3x+2)$.

6 Bob the trickster played a number trick with Duke the dimwitted. He told him to pick an even number, double it, add 48, divide by 4, subtract 7, multiply by 2, and subtract his original number. Bob then told him the result he should have attained. Duke was surprised and chose a different number to do it again. He got the same answer. What number did Bob tell Duke to surprise him?

7 Factor each of the following expressions.

(a) $6a^2 - 24a$

(b) $4x^2 - 18x$

(c) $-8t^2 - 4t$

(d) $6x^3 - 12x^2 + 12$

8 Factor the expression $4x(x-2) + 7(x-2)$.

In the last part of this book, you will learn about how to simplify rational expressions and how to solve rational equations. First, when we see rational expressions, the key idea is to make *common denominator*.

9 Write each of the following expressions as a single fraction.

(a) $\dfrac{2}{x} + \dfrac{3x}{4}$

(b) $\dfrac{4}{3x} - \dfrac{5-x}{6x^2}$

10 Write

$$\dfrac{2}{x} + \dfrac{7x}{x+3}$$

in a single fraction.

1 Evaluate the following expressions when $x = 2$.

(a) $x^2 - 4$

(b) $\dfrac{7x}{10} + \dfrac{3x}{10}$

(c) $(6x - 1)(2x - 3)$

(d) 2^{2x}

2 Simplify $(2x^2)(5x^4)$.

3 Simplify the following expressions.

(a) $\dfrac{(-8a^4)(4a^3)}{(12a^2)(2a^3)}$

(b) $\dfrac{3x^2}{2x^8} \times \dfrac{6x^4}{5x^2} \times \dfrac{10x^8}{3x^2}$

4 Expand $4y\left(\dfrac{y}{4} + \dfrac{4}{y} + \dfrac{3}{4y}\right)$.

5 Expand the expression $4(x^2 - 4x + 1) - x(x - 2)$.

6 Factor the following expressions.

(a) $2y(y^2 + 1) + 5(y^2 + 1)$

(b) $2(4d + 3) + 3d(4d + 3)$

7 Write the following expressions into a single fraction.

(a) $\dfrac{1}{2x^2} - \dfrac{2 - x}{4x^3}$

(b) $\dfrac{x^2 - x}{x^2 - 1} + \dfrac{3x^3}{15x^4}$

1

(a) 0 (b) 2 (c) 11 (d) 16

2 $10x^6$

3

(a) $-\dfrac{4}{3}a^2$ (b) $6x^2$

4 $y^2 + 19$

5 $3x^2 - 14x + 4$

6

(a) $(2y+5)(y^2+1)$ (b) $(2+3d)(4d+3)$

7

(a) $\dfrac{3x-2}{4x^3}$ (b) $\dfrac{5x^2+x+1}{5x^2+5x}$

Topic 2

More Variables

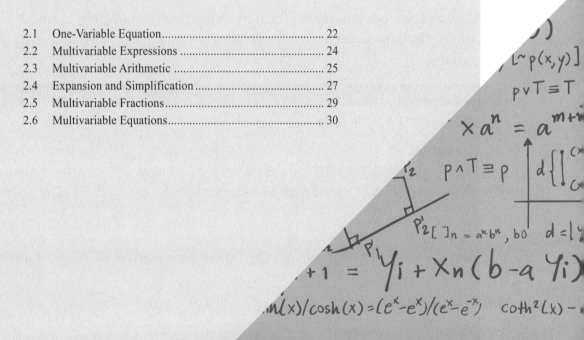

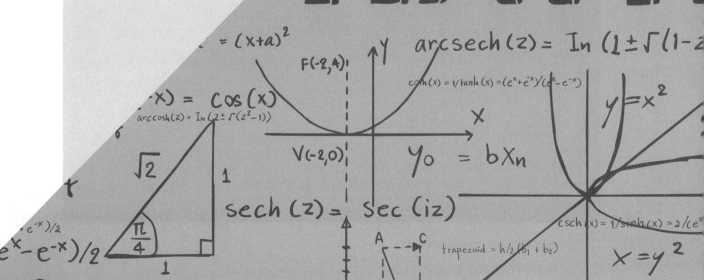

2.1 One-Variable Equation

In Prealgebra, we learned how to solve $ax + b = c$ by isolating the variable. "Solving" the equation means finding the correct value of x for the given equation. For instance, given $3x + 5 = 2x + 7$, we can solve it by

$$3x + 5 = 2x + 7$$
$$3x - 2x = 7 - 5$$
$$x = 2$$

In Algebra I, we complicate the form by extending the meaning of linear equations. As long as we have $a \bigcirc + b = c \bigcirc + d$, we can solve for $\bigcirc$, by isolating the variable.

Example

Solve for x if $5x - 3 = 2x + 9$.

Solution
We isolate the variable by subtracting $2x$ to both sides of the equation. Then, $3x - 3 = 9$. Hence, add 3 to both sides to get $3x = 12$. Thus, $x = 4$.

$\boxed{1}$ Find the value of x that satisfies the following equation.

(a) $5\sqrt{x} - 2 = 28 - \sqrt{x}$

(b) $\dfrac{6}{x} + 3 = 7 - \dfrac{2}{x}$

2 Solve the following equations.

(a) $\sqrt[3]{1-2t}+3+2\sqrt[3]{1-2t}=6$

(b) $\dfrac{x}{x+1}+\dfrac{3}{4}=\dfrac{3}{x+1}$

3 Find all possible values of x for

$$(x-3)^2+4=6-(x-3)^2$$

2.2 Multivariable Expressions

Just as we evaluate one-variable expression by substituting the x-value into the expression, we reiterate the same process for multivariable expressions.

Example

Evaluate $2xy - \dfrac{3y}{x^2}$ if $x = 2$ and $y = 4$.

Solution

Substituting $x = 2$ and $y = 4$ into the original equation, we get

$$2xy - \frac{3y}{x^2} = 2(2)(4) - \frac{3(4)}{2^2}$$
$$= 16 - 3$$
$$= 13$$

4 Evaluate each of the following expressions when $r = 3$ and $s = -2$.

(a) $r^2 + 2rs + s^2$

(b) $(r + s)^2$

(c) $r^3 + 3r^2s + 3rs^2 + s^3$

(d) $(r + s)^3$

2.3 Multivariable Arithmetic

Example

Simplify $(2x+3y+4)-(5x-11y)$.

Solution

Group similar expressions together.

$$(2x+3y+4)-(5x-11y) = (2x-5x)+(3y-(-11y))+4$$
$$= -3x+14y+4$$

5 Simplify each of the following expressions.

(a) $(x+y-z)+(2x-3y+11z)$

(b) $4ab+3cd+2cd-11ab+4-1$

6 Simplify the product $(3xy^2) \cdot (2xy^6)$.

$\boxed{7}$ Simplify each of the following expressions.

(a) $\dfrac{28x^3y^2}{21x^4y^6}$

(b) $\dfrac{-4x^2y^3z^7}{-16x^3y^7z^2}$

$\boxed{8}$ By what expression can we multiply $6x^2y^3$ to get $18x^5y^6$?

2.4 Expansion and Simplification

Just as we expand or simplify one-variable expression, we can perform similar expansion or simplification with more variables.

Example

Expand the product $xy(x-y)$.

Solution
Using the distributive property, we get

$$xy(x-y) = (xy)x - (xy)y$$
$$= x^2y - xy^2$$

9 Expand the product $(a+b)(2a+3b)$.

10 Simplify the following expressions.

(a) $(2a+3b) - (4a+6b)$

(b) $4(x-y+z) - 3(2x-2y+2z)$

Factor $25ab + 30bc$.

Solution

#1. Find out the common number and common variable.

#2. Factor the expression with the common divisor.

$$25ab + 30bc = 5b(5a + 6c)$$

11 Completely factor the following expressions.

(a) $6x^2 + 8xz$

(b) $7a^2b^2 - 21ab^3 + 14a^2b^3$

12 Simplify the product $\dfrac{2x + 4y}{8} \cdot \dfrac{3xy}{x^2 + 2xy}$.

2.5 Multivariable Fractions

Fractions with multivariable expressions can turn into a single fraction with a common denominator.

Example

Write $\dfrac{3}{a} - \dfrac{2}{b}$ as a single fraction by finding a common denominator.

Solution

$$\frac{3}{a} - \frac{2}{b} = \frac{3b}{ab} - \frac{2a}{ab}$$
$$= \frac{3b - 2a}{ab}$$

13 Simplify the following multivariable expressions into single fractions.

(a) $\dfrac{5y}{6x^2} - \dfrac{4}{3xy}$

(b) $\dfrac{2a^3}{a^3b} + \dfrac{3b}{a} - \dfrac{3b - 3}{6ab - 6a}$

2.6 Multivariable Equations

Just as expressions can have more than one variable, so can equations, such as

$$x + 2y = 3.$$

Just as we can isolate the variable in some one-variable equations, we can sometimes isolate one variable in an equation with multiple variables. For example, we isolate x in the equation $x + 2y = 3$ by subtracting $2y$ from both sides, which gives

$$x = 3 - 2y.$$

We call this "solving the equation for x in terms of y." We can also solve $x + 2y = 3$ for y in terms of x by first subtracting x from both sides of $x + 2y = 3$ to get $2y = 3 - x$, then dividing by 2 to find

$$y = \frac{3 - x}{2}.$$

We would like to see a special type of multivariable equations, which involves "integers" and two variables.

$\boxed{14}$ If $xy - 2x - 3y = 2$ for positive integers (x, y), find all possible solution pairs of (x, y).

1. Evaluate each of the following expressions when $a = -3$ and $b = 4$.

(a) $ab + 2b + 3a$

(b) $\dfrac{a^2}{b^3}$

(c) $4ab(a + b)$

(d) $\sqrt{-ab} \times \sqrt{3}$

2. Find the value of y if $4x + 3y - 2x + 4y = 1$ for $x = -2y$.

3 Expand the following expressions.

(a) $(2x - 7y)(3x - 2y)$

(b) $2ab(3 + 2b) + 3a(b - 4b^2)$

4 Simplify the following expressions. Assume all variables are positive.

(a) $(-2xy)^4 + (4xy^3)(-3x^3y)$

(b) $\sqrt{ab^2}\sqrt{a^3b^4}$

5 Evaluate

$$3(x^2 - 2x + 1) + 4(y^2 - 3y + 2)$$

when $2x = y$ and $x = 1$.

6 Factorize $24x^3yz^2 + 18x^2y^2z^3$ where x, y, and z are positive real numbers.

7 Turn $\dfrac{1}{a} + \dfrac{1}{b} + \dfrac{1}{c}$ into a single fraction in the lowest term.

8 Express $\dfrac{x^2 - 2x}{3x - 6} + \dfrac{4y^2 - 3y}{8y - 6}$ as a single fraction.

9 Find all (x, y, z) satisfying the equation $3x + 2y + z = 10$ where x, y, and z are positive integers.

1

(a) -13 (b) $\dfrac{9}{64}$ (c) -48 (d) 6

2 $y = \dfrac{1}{3}$

3

(a) $6x^2 - 25xy + 14y^2$ (b) $9ab - 8ab^2$

4

(a) $4x^4y^4$ (b) a^2b^3

5 0

6 $6x^2yz^2(4x + 3yz)$

7 $\dfrac{ab + bc + ca}{abc}$

8 $\dfrac{2x + 3y}{6}$

9 $(x, y, z) = (1, 1, 5), (1, 2, 3), (1, 3, 1), (2, 1, 2).$

Topic 3

Two-Variable Equations

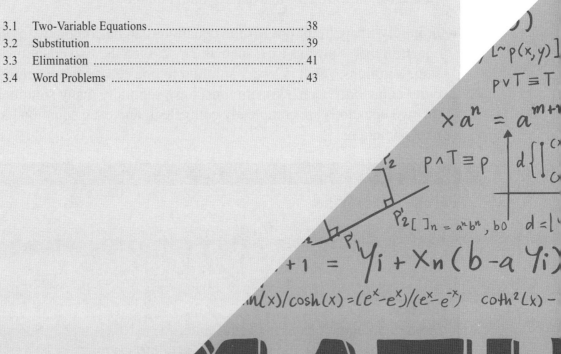

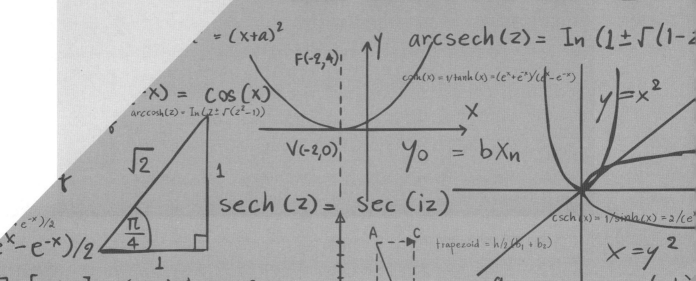

3.1 Two-Variable Equations

There are three possibilities for the number of solutions of a system that consists of a pair of two-variable linear equations:

- No Solution : If the two equations can be combined to produce an equation that is never true, such as $3 = 4$, then there is no solution to the system of equations.

- One Solution : In this case, the equations can be manipulated to find unique values for the variables that satisfy both equations.

- Infinitely Many Solutions : If the equations can be combined to produce an equation that is always true, such as $2 = 2$, then the two equations have the same solutions. (That is, every solution to one equation is a solution to the other equation.) Such a system must have infinitely many solutions because the two equations are completely equivalent, and each equation has infinitely many solutions.

Example

Find all (x, y) such that

$$3x - 2y = 4$$
$$2x + 2y = 6$$

Solution
Adding two equations at the same time, $5x = 10$, so $x = 2$. Substituting $x = 2$ into the first equation, $3(2) - 2y = 4$ implies that $y = 1$.

1 Find three different ordered pairs (x, y) that satisfy both of the equations below.

$$2x - 3y = 1$$
$$6x - 9y = 3$$

3.2 Substitution

In order to solve the system of equations, we use the technique of *substitution*, i.e., substituting one variable expression into another expression to reduce the system to one-variable equation.

Example

Solve the following system of equations.

$$2x + \frac{y}{2} = \frac{17}{2}$$
$$-\frac{x}{2} + 2y = 17$$

Solution

Since $2x + \frac{y}{2} = \frac{17}{2}$, then $y = 17 - 4x$. Substituting this expression into the second equation, $-\frac{x}{2} + 2(17 - 4x) = 17$, so $x = 2$. Hence, $y = 9$.

$\boxed{2}$ Solve the following system of equations.

$$0.3x - 0.5y = 0.1$$
$$2x - 0.7y = 3.3$$

3 Solve the following system of equations for p and q.

$$3p - q = 14$$
$$2p + 3q = -20$$

4 Solve the following system of equations for x and y.

$$2x - 2y = 1000$$
$$x + 2y = 1019$$

3.3 Elimination

The second method of solving the system of equations is *elimination*. We eliminate one variable by properly adding or subtracting two equations.

Example

Solve for x and y.

$$5x + 2y = -11$$
$$3x - 2y = 67$$

Solution

Adding two equations at the same time, $8x = 56$, so $x = 7$. Hence,

$$5(7) + 2y = -11$$
$$35 + 2y = -11$$
$$2y = -46$$
$$y = -23$$

Therefore, $(x, y) = (7, -23)$.

$\boxed{5}$ Solve the system of equations for x and y.

$$4x - 3y = 18$$
$$y + 2x = -3y - 2$$

The process of elimination or substitution can be extended to the system of linear equations with more than two variables.

6 Solve the following system of linear equations with three variables.

$$a + 2b + 3c = 14$$
$$2a - b + 4c = 12$$
$$5a + 3b - 6c = -7$$

3.4 Word Problems

Example

A football game was played between two teams, Chelsea and Liverpool. The two teams scored a total of 6 points, and Chelsea won by a margin of 2 points. How many points did Liverpool score?

Solution
Let C, L be the scores each team scored respectively. Then,

$$C + L = 6$$
$$C - L = 2$$

Thus, $2C = 8$, so $C = 4$, and $L = 2$.

7 Bob has only nickels and quarters in his pocket. Their combined value is \$4.75. If he has 5 more nickels than quarters, how many quarters does he have in the pocket?

8 Two years ago, Amy was nine times as old as Bob. Amy is now seven times as old as Bob is. Find the sum of their ages now.

9 The sum of Eric's and Ewan's weights is 8 times the difference of their weights. The positive difference of their weights is also 50 kilograms less than the sum. If Eric weighs less than Ewan, find the sum of their weights, and Eric's weight.

10 If

$$5g + 2r = 10$$
$$g + 4r = 7$$

find the value of $12g + 12r$, for (g, r) satisfying the given system of equations.

1 Solve the following systems of equations by substitution.

(a) $\begin{cases} x + 4y = -5 \\ 2x - 8y = 70 \end{cases}$

(b) $\begin{cases} 3x - y = 4 \\ 6x + 4y = 20 \end{cases}$

2 Solve the following systems of equations by elimination.

(a) $\begin{cases} 4x - 6y = -34 \\ 7x + 3y = 35 \end{cases}$

(b) $\begin{cases} 2x + 8y = 12 \\ -6x + 16y = 4 \end{cases}$

3 If $\begin{cases} 2x + y = 1 \\ 2y + x = 0 \end{cases}$, then find the sum $x + y$.

4 Bob's age 5 years ago plus twice Amy's age now gives 60, while Amy's age 5 years from now plus twice Bob's age now gives 120. What is the sum of their ages now?

5 Find the number of ordered pairs (x,y) such that

$$3 + \frac{x}{y} = \frac{2}{y}$$
$$3x + 9y = -4$$

6 If

$$x + y = 6$$
$$y + z = 10$$
$$z + x = 12$$

find the ordered triple (x,y,z) satisfying the system of equations above.

1

(a) $(x, y) = (15, -5)$ (b) $(x, y) = (2, 2)$

2

(a) $(x, y) = (2, 7)$ (b) $(x, y) = (2, 1)$

3 $x + y = \dfrac{1}{3}$.

4 The sum of their ages is 60.

5 There is no ordered pair (x, y) satisfying the system of equations.

6 $(x, y, z) = (4, 2, 8)$.

Topic 4
Proportion

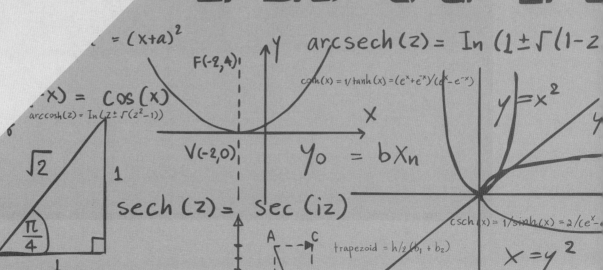

4.1 Direct Proportion

If x and y are directly proportional, then the quotient x/y (or y/x) stays constant. In other words, $x/y = k$ for some real constant k. Another way to say this is to write an expression such as $x = ky$.

Example

A secret recipe requires 4 cups of sugar and 3 cups of flour. You only have 2 cups of flour. How many cups of sugar do you need if all 2 cups of flour are used?

Solution

Set up a direct proportion equation such that $\dfrac{4}{3} = \dfrac{x}{2}$. Hence, $3x = 8$, so $x = \dfrac{8}{3}$ cups of sugar are needed.

Shadow length type is an application of direct proportion. The length of an object's shadow is directly proportional to its height, assuming that everything is located in the same position at the same time of the day.

1 Bob is 5 feet tall whose shadow is 10 feet long. The flagpole he is standing next to casts a shadow that is 32 feet long. Assuming that the shadow length is proportional to the height of the given object, find the length of the flagpole.

4.2 Inverse Proportion

If x and y are inversely proportional, then the product xy stays constant. We can write this as $xy = k$, where k is a real number, sometimes called as the constant of variation(or proportionality). Typical examples of inverse variations are applications of work rate question.

Example

If a and b are inversely proportional and $a = 3$ when $b = 8$, find the value of a if $b = 12$.

Solution

Since $ab = k$ for some constant k, then, $ab = (3)(8) = 24$ by substituting the given values. Therefore, $a \cdot (12) = 24$, so $a = 2$.

$\boxed{2}$ Twelve people together can paint a room in eighteen hours. In how many hours could nine people have painted the same room? (Assume that all people paint a room at the same rate.)

$\boxed{3}$ Suppose that x varies inversely with y^2. If $x = 9$ when $y = 2$, find the value of x when $y = 10$.

4.3   Joint Proportion

We say a variable x is jointly proportional to y and z if

$$x = kyz$$

Example

a is jointly proportional to $\sqrt{b}$ and c^2, and $a = 3$ if $b = 4$ and $c = 3$. Find the constant of variation.

Solution
Since $a = k\sqrt{b}c^2$,

$$a = k\sqrt{b}c^2$$
$$\frac{a}{\sqrt{b}c^2} = k$$
$$\frac{3}{\sqrt{4}3^2} = k$$
$$\frac{1}{6} = k$$

4 Suppose a is jointly proportional to b and c. If $a = 8$ when $b = 4$ and $c = 12$, then what is the value of a when $b = 4$ and $c = 9$?

Example

If Bob reads a page in 1 minute, how many pages of a book can he read in 1 hour?

Solution
Let's set up a proportion equation.

$$1 \text{ hour} \times \frac{60 \text{ minutes}}{1 \text{ hour}} \times \frac{1 \text{ page}}{1 \text{ minute}} = 60 \text{ pages}$$

5 Amy and Bob read a 700-page novel. Amy reads one page every minute, while Bob reads one page every 40 seconds. Amy starts reading at 1:00, and Bob starts reading 30 minutes later. When will Bob catch up to Amy's page?

6 Amy and Bob runs 10 kilometers per hour towards each other, both of whom are initially 60 kilometers apart. A dog starts at the front of Amy, runs to Bob at a straight line, then back to Amy, and so on. If the dog always runs at 30 kilometers per hour, how far does the dog run before Amy and Bob catch up with one another?

Suppose Bob is at the airport. If he walks in a sidewalk that moves 1 meter per second, he can reach Gate #5 in 1 minute, which is 200 meters away from the security check. Assume that the side walk starts right off the security check. How fast is he, if he walks off the sidewalk by himself?

Solution

Let x be his speed in meter per second. Then,

$$\frac{x+1 \text{ meters}}{1 \text{ second}} \times 60 \text{ seconds} = 200 \text{ meters}$$

Hence, $x+1 = \dfrac{10}{3}$. Therefore, $x = \dfrac{7}{3}$ meters per second.

---

7 Bob is filling a water tank that is initially empty. It ordinarily takes him 50 minutes to fill the whole tank. Charlie, on the other hand, empties the tank while Bob tries to fill it. It would take 70 minutes for Charlie to empty the full tank by himself. If they work simultaneously, how long will it be until the whole tank is filled?

Suppose Bob is on the boat. The speed of the boat in still water has the speed of 2 meters per second. However, the river in which his boat currently stays flows downstream with the constant speed, the numerical value of which is unknown to Bob. Fortunately, he takes a timer out to calculate the time it takes his boat to travel either upstream or downstream. Specifically, it takes 5 seconds to travel 15 meters downstream. On the other hand, it takes 15 seconds for Bob to return to the original point. Find the speed of the current.

Solution

Let x be the speed of the current in meter per second.

$$\begin{cases} (2+x)5 = 15 \\ (2-x)(15) = 15 \end{cases}$$

Hence, $2+x = 3$ and $2-x = 1$. Therefore, $x = 1$ meter per second.

8 A boat travels between City A and City B, which are 1,000 kilometers apart. City A is due north of City B, and there is a strong, constant water current flowing due north. At a constant speed, it takes the boat 5 hours to go from B to A with the help of the current, but 8 hours to go back when fighting against the current. What is the speed of the water current?

1 Suppose a and b are directly proportional. If $a = 10$ when $b = 24$, what is a when $b = 6$?

2 Suppose x is inversely proportional to $\sqrt{y}$. If x is quadrupled, what happens to y?

3 If x^2 and y^3 are directly proportional, and $x = 2$ when $y = 3$, then what is y when $x = 3$?

4 Bob rides his bike to work, whose office is 10 kilometers from his home, and he rides 12 kilometers per hour. At what time should he leave to get to work at 9 : 00 A.M.?

5 Suppose x is directly proportional to y, but inversely proportional to z. If $x = 3$ when $y = 6$ and $z = 12$, then what is the value of z when $y = 8$?

6 Jack began placing a pile of 500 apples into a box at the rate of 10 apples per minute. Five minutes later, Dylan joined him and placed fruits(=apples, in this case) into the box at the rate of 5 apples per minute. Assuming that their work rate did not change while they worked, how many apples had Dylan moved on his own when they finished moving the fruits?

7 Amy and Bob, who are professional runners, stood on a circular track 400 meters around. Bob started running at a rate of 10 meters per second. Amy waited for a full minute, then began running from the same point (in the same direction) at 15 meters per second. How many seconds should have elapsed before Amy physically passed Bob on the track for the first time?

8 Amy and Bob, who are professional architects, are building a fence. Amy can build the fence alone in 4 hours. If Bob starts helping Amy after she has already worked on the fence for 2 hours, they will finish the fence 1 hour and 30 minutes right after he joins her. If their work rate stays constant the whole time they work, how long would it take, in average, for Bob to build the fence alone?

1 $\dfrac{5}{2}$

2 y is divided by 16.

3 $y = 3\sqrt[3]{\dfrac{9}{4}}.$

4 8 : 10 A.M.

5 $z = 9$

6 150 apples

7 40 seconds

8 12 hours

Topic 5

Linear Graphs

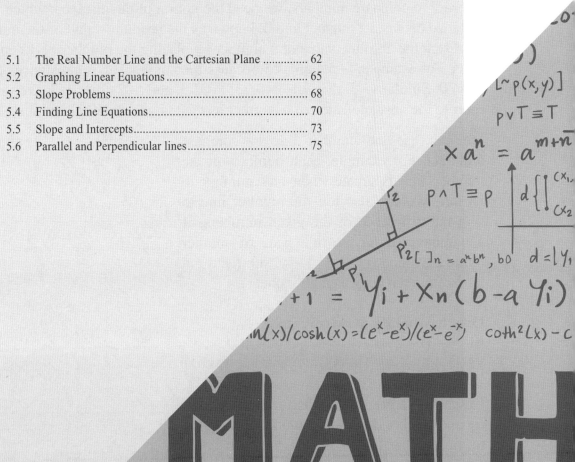

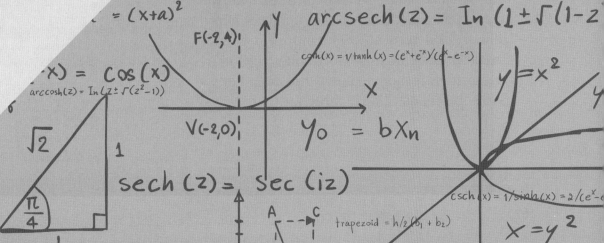

Real numbers are extremely useful in terms of comparison. We can even plot the numbers on a straight line, called the number line.

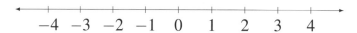

If you see that two numbers are real, it means that we can COMPARE the numbers. In order to compare numbers, we need the notion of the distance, which is the distance between a number and 0, known as the absolute value of the number. We represent the absolute value of a number by placing the number between vertical lines. For example, $|-5|$ equals 5 because the number -5 is a distance of 5 away from 0. Similarly, $|3.1| = 3.1$, $|-0.73| = 0.73$, and $|-2/3| = 2/3$. We sometimes refer to the absolute value of a number as the magnitude of the number.

We can use the number line to compare numbers. For example, each number on the number line is greater than any number to its left. We can also use the number line to find the difference of the given numbers. If a number x is on the right side of another number y, then $|x - y| = x - y$, subtracting a larger number(y) by a smaller one(x).

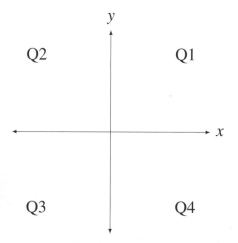

Let's have a look at two examples.
#1. $|(-3) - 2| = 2 - (-3) = 5$.
#2. $|1 - 4| = 4 - 1 = 3$.

Descartes' great insight that unified geometry and algebra was adding another dimension to the number line. This insight has become so important we still use Descartes' name to describe the result, a.k.a. Cartesian plane. Instead of just plotting numbers horizontally, on the Cartesian plane we plot numbers horizontally and vertically. The introduction of Cartesian plane helps us solve plane geometry questions in more intuitive direction. Let's have a look at question number 1.

1 Plot three points on the Cartesian plane below : $(3,2)$, $(-2,-2)$ and $(0,-3)$.

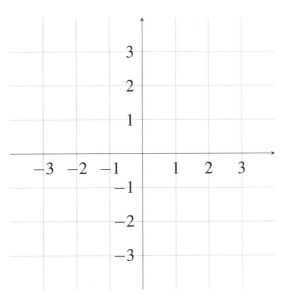

2 Find the distance between $(-2,-3)$ and $(3,2)$.

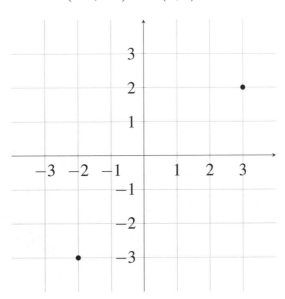

3 Find the distance between the points $(-3,4)$ and $(5,4)$.

4 Determine which of the following points is farthest from the origin.

(A) $(0,4)$ (B) $(1,2)$ (C) $(4,3)$ (D) $(2,4)$

5.2   Graphing Linear Equations

The graph of a linear equation looks like a line. Here, a line can be plotted by connecting any two points on the graph and extending the segment in proper directions. Interestingly, the ratio of the difference in y-values to the difference in x-values does not change. Let's have a look at this property in the following questions.

$\boxed{5}$ Plot several solutions to the equation $x - y = 1$. Plot the solutions (x, y) on the cartesian plane.

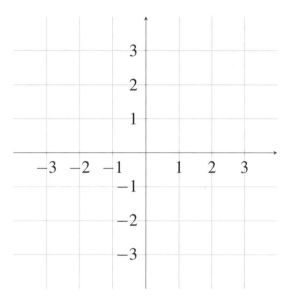

$\boxed{6}$ Find several solutions to the equation $x + 2y = 12$. For each pair of solutions, find the ratio of the difference in y values to the difference in x values.

The ratio found in question 6 is called the slope of a line. The slope of the line, as illustrated before, is computed by

$$\frac{\triangle y}{\triangle x} = \frac{y_2 - y_1}{x_2 - x_1}$$

7 Given two points (x_1, y_1) and (x_2, y_2), when does a line have a negative slope? How does it appear in a line's graph if the line has a negative slope?

8 Find the slope of the line that passes through $(-3, 5)$ and $(2, -5)$.

9 Graph the following equations.

(a) $y = -1$.   (b) $x = 1$.

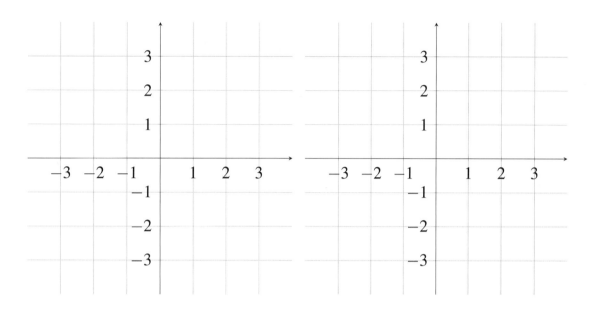

10 For the following graph, determine whether the slope is positive, negative, or 0.

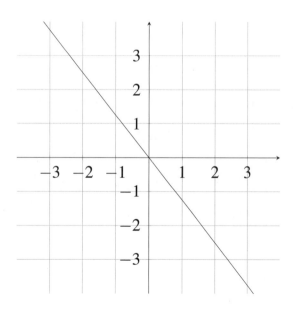

5.3 Slope Problems

Given a slope m and the point (h,k), we can always plot the equation in the Cartesian plane.

#1. First, plot the point.

#2. Second, use the slope to find out other points.

11 Graph the line that passes through the point $(-3,2)$ and has slope of $-\dfrac{1}{3}$.

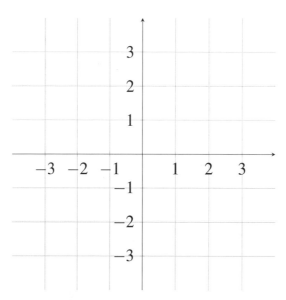

Three points A, B, and C are collinear if $\overline{AB}$ and $\overline{BC}$ have the same slope values.

12 Determine whether the three points lie on the same straight line.

$A(32,5)$ $B(24,18)$ $C(22,21)$

13 Let point P be $(2,4)$ and Q be $(-7,-2)$.

(a) What are the coordinates of the midpoint[1] of segment $\overline{PQ}$? Explain why this point must be the midpoint of $\overline{PQ}$.

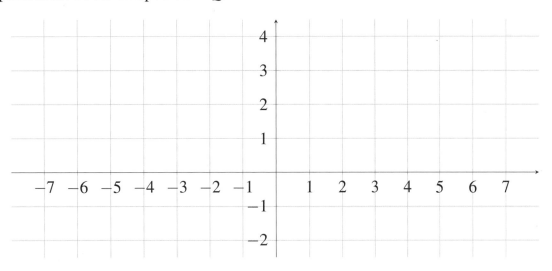

(b) Let T be the point on segment $\overline{PQ}$ such that $PT:TQ=1:2$. Find the coordinates of point T, and explain why this point T must be the point on $\overline{PQ}$ such that $PT:TQ=1:2$.

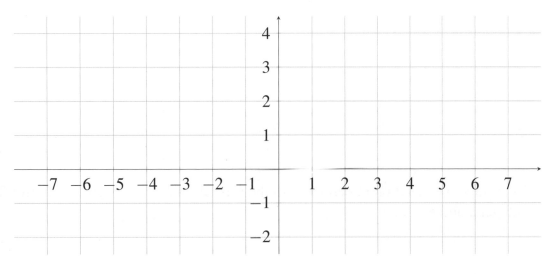

---

[1]We denote the line that passes through points A and B as $\overleftrightarrow{AB}$. Also, the point on $\overline{AB}$ that is the same distance from A as it is from B is called the midpoint of the segment $\overline{AB}$.

5.4   Finding Line Equations

Let's use the fact that the slope of a line does not change for any point (x, y) on the line in order to find out the line equation.

14

(a) Find the coordinates of any three points on the line.

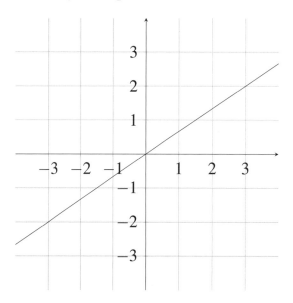

(b) Let the point (x, y) be on the line other than $(3, 2)$. In terms of x and y, what is the slope between (x, y) and $(3, 2)$?

(c) Rearrange the equation in part (b) into the form $Ax + By = C$ where A, B, and C are integers, and $A > 0$.

15 Suppose the line passes through $(0,4)$ and $(5,-3)$.

(a) Find the slope of the line.

(b) Find the line equation passing through these two points.

16 Find the equation of the following lines such that

(a) it passes through $(3,1)$ with slope -2.

(b) it passes through $(3,2)$ and $(-5,2)$.

17 Consider all points whose coordinates are $(x(t), y(t)) = (4t + 2, 3t - 2)$ for some real number t. Find the line equation in standard form.

18 Let A be $(3, 7)$, B be $(-1, -5)$, and C be $(5, 3)$. The median of a triangle connects a vertex of a triangle to the midpoint of the opposite side. Find an equation describing the line that contains the median from B to the midpoint of $\overline{AC}$.

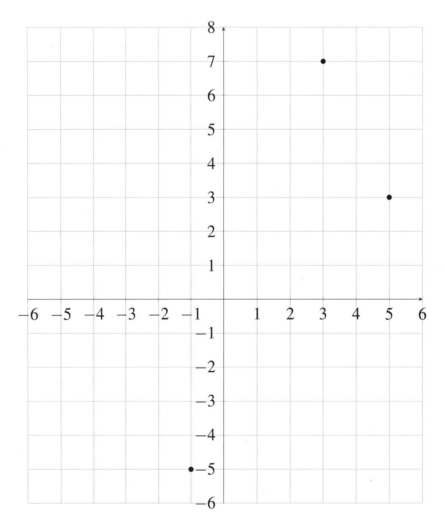

5.5 Slope and Intercepts

In the previous sections, we learned that two points on the line can determine the line equation. More importantly, the points where a line meets either one of the axes can be particularly useful. These points are called the intercepts of the line. Any point where a graph hits the x-axis is an x-intercept of the graph and any point where a graph hits the y-axis is a y-intercept. Use the following set of rules to find the intercepts.

- y-intercept : set $x = 0$.

- x-intercept : set $y = 0$.

$\boxed{19}$ Given a line equation $y = 1 - \dfrac{1}{2}x$,

(a) find two points on the line.   (b) find the slope of the line.

(c) graph the line.

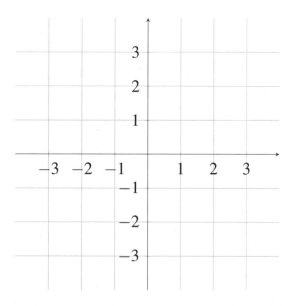

20 A line with the slope value of 2 intersects a line with the slope of 5 at the point $(10, 20)$. What is the area of the triangle formed by the two lines and the x-axis?

21 Bob begins jogging at a steady rate of 3 meters per second down the middle of lane one of a straight track. Amy starts even with him in the center of lane two but moves at 4 meters per second. At the instant they begin, Charlie is located 100 meters down the track in lane three, and is heading towards them in his lane at 6 meters per second. After how many seconds will the runners lie in a straight line?[2]

---

[2]One key step was writing all the x-coordinates in terms of the same variable, t. The process of expressing multiple quantities in terms of the same variable is called parametrization. The common variable is called the parameter.

5.6 Parallel and Perpendicular lines

Two lines in a plane that never intersect are called parallel lines, while two lines in the same plane that form right angles when they intersect are called perpendicular lines. Below is an example of a pair of parallel lines, and a pair of perpendicular lines.

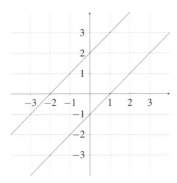

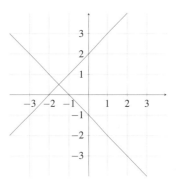

Parallel lines   Perpendicular lines

If two lines meet at a point, just as in the example of perpendicular lines, we call that point as the point of intersection, which happens to be the solution of the system of equations at the same time.

22 On the same Cartesian plane, graph both the equations in the system of equations.

$$4x + 2y = 8$$
$$2x + 3y = 6$$

Use your graph to find the solution to this system of equations.

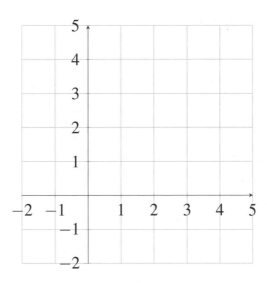

23 Solve the following questions.

(a) Find an equation of the line through $(2, 1)$ parallel to the graph of $2x - 7y = 11$.

(b) What is the x-intercept of the line through $(2, -2)$ that is perpendicular to the graph of $x - 2y = 5$?

24 The graphs of the equations $3x - By = 14$ and $Ax + 3y = 21$ are both the same line. Find A and B.

1 Suppose there is a line equation that passes through $(0,6)$ and $(3,0)$. Find the x-coordinate of the point on the graph whose y-coordinate is 5.

2 Given two real numbers a and b, $|a-b|$ is the distance between a and b. Find the largest of the two numbers and subtract it by the smaller one. Hence, compute

$$|2019 - |2018 - |2017 - |2016 - |2015|||||$$

3 Consider the points $(3,2)$ and $(-3,-3)$. Plot the points in the following grids and find the distance between the two points.

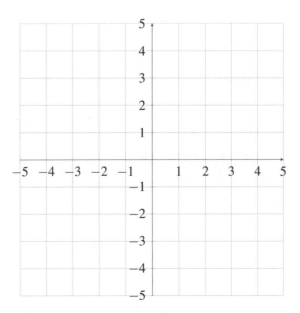

4 A is a constant such that the graph of the equation $Ax + 2y = 6$ passes through the point $(-2, 4)$. Find the value of A. Hence, find the area of the triangle whose vertices are the x-intercept, y-intercept, and the origin.

5 The lines l, p, and q are graphed in the following graphs. The line l has the slope of $-3/4$, p has the slope of $1/6$, and q has the slope of 2. Label the proper letters to the following graphs without finding the coordinates of any points on any of the lines.

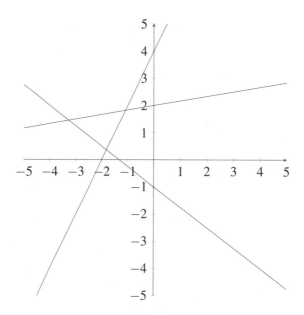

6 If there is a line that passes through $(3, 1)$ with the slope of $\dfrac{2}{5}$, find the equation and graph the line on the following diagram.

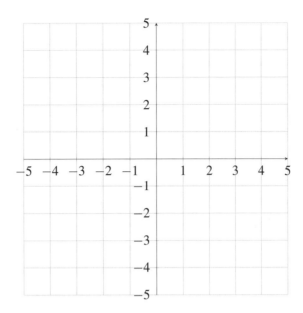

7 Given two distinct points, there is a unique line passing through the two points in the plane. Given two points $(2, -6)$ and $(-1, 3)$, find the x-intercept and the y-intercept of the line that passes through the two.

8 Given the linear equation $2x + 3y = 6$, graph the equation in the following diagram, and find its slope, x-intercept, and y-intercept.

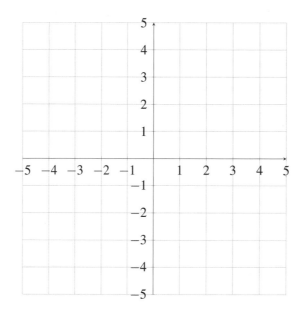

9 Suppose $P(-3, 10)$, T and $Q(6, -14)$ are collinear, where $PT : TQ = 1 : 2$, find the exact coordinates of T. (Assume that T is between P and Q.)

1 $x = \dfrac{1}{2}$

2 2017

3 The distance between the two points is $\sqrt{61}$.

4 $A = 1$, and the area of triangle is 9.

5 l is the line with the negative slope, so the graph must be falling. p is the line with the wide positive slope, and q is the line with steep positive slope.

6 The equation is $y = \dfrac{2}{5}x - \dfrac{1}{5}$, and the graph can be given by

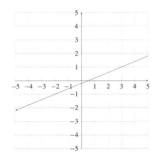

7 The equation is $y = -3x$. Hence, the x-intercept is 0, and the y-intercept is 0.

8 The slope is $-\dfrac{2}{3}$, x-intercept is 3 and y-intercept is 2. The graph can be given by

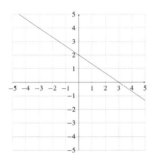

9 The exact coordinates of T should be $(0, 2)$.

Topic 6

Basic Inequalities

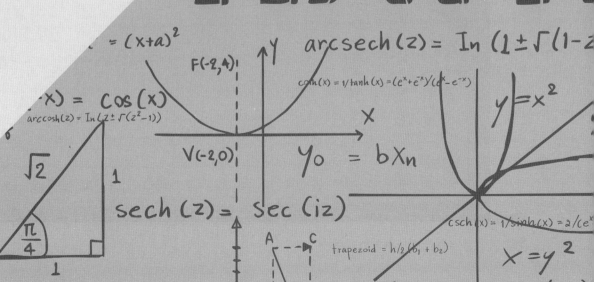

Now, the main property of real numbers is that we can COMPARE the numbers. So far, we use the equation to solve for a certain variable or simplify the expressions. In this topic, we will look at $a \neq b$ resulting in two cases when the left-side of the equation is different from the right side of it.

$$a > b \qquad a < b$$

Both of the inequalities above are called strict inequalities. That being said, we can also write less-strict inequalities, in which one side is greater than or equal to the other. For example, because $3 - 1$ is "greater than or equal to" 2, we can write

$$3 - 1 \geq 2.$$

- If $x > y$ and $y > z$, then $x > z$.

- If $x > y$, then $x \pm a > y \pm a$.

- If $x > y > 0$ and $a > b > 0$, then $xa > yb$.

- If $x > y$ and $a < 0$, then $ax < ay$.

- If $x > y$ and x and y have the same sign, then $\dfrac{1}{x} < \dfrac{1}{y}$.

$\boxed{1}$ If $a \leq b$ and $a \geq b$, then $(a+b)(a-b) = ?$

$\boxed{2}$ Solve the following questions.

(a) Compare the following two numbers.

$$2^{-3} \qquad 3^{-3}$$

(b) Compare the following two numbers.

$$\frac{12}{5} \qquad \frac{9}{7}$$

(c) Suppose Jeff Bezos has more assets(or money) than Bill Gates. If they both donate 1 billion dollars for social goods, then who will have more assets?

$\boxed{3}$ Is a square of a real number positive, negative, or 0?

6.2 Comparison

The trichotomy of real numbers tells us that for any two numbers a and b, either one of the following is true. The idea of comparison makes us think about three cases all the time. This means that comparing two real numbers makes us perform caseworks.

$$a > b \qquad a = b \qquad a < b$$

How do we compare two numbers if they are written in exponential expressions? We usually try one of the two strategies.

- Equalize the bases in prime factorization, if possible.

- Equalize the exponents by finding their least common multiple.

4 Compare the following numbers and list from least to greatest.

$$2^{100} \qquad 3^{80} \qquad 5^{40}$$

5 Determine which of

$$A = \frac{54321}{54322} \qquad \text{and} \qquad B = \frac{5432}{5433}$$

is larger without using a calculator.

6 Which number is larger?

$$\sqrt{4\sqrt{3}} \qquad \sqrt{2\sqrt{5}}$$

7 Which number is larger?

$$\frac{1}{1+2+3+4+\cdots+2018+2019} \qquad \frac{1}{2+3+4+5+\cdots+2019+2020}$$

8 First, compare 2^{80} and 3^{50}. Then, determine which of the two numbers - 2^{82} and 3^{48} - is larger.

6.3   Linear Inequalities

Let's have a look at linear inequality involving a variable expression such as $x > 8$. This inequality tells us that x is greater than 8, meaning that x can be 10, but not 5. We can graph the solutions to $x > 8$ on the number line.

Solutions to inequality problems are sometimes written using interval notation. For example, we can denote 'all numbers greater than or equal to -3 and less than or equal to 5' by the interval $[-3, 5]$. We use '[' and ']' to indicate that the values -3 and 5 are included. Notice that we don't write $[5, -3]$; the smaller number between the two endpoints always comes first. To use this notation to indicate $-3 \le x \le 5$, we write $x \in [-3, 5]$, where '$x \in$' means 'x is in.'

To exclude a boundary value from an interval, we use '(' for the lower bound and ')' for the upper bound. For example, the statement $x \in (-3, 5]$ means $-3 < x \le 5$ and $y \in (-12.2, 0)$ means $-12.2 < y < 0$. Finally, to indicate an interval that has either no upper bound or no lower bound (or neither), we use the ∞ symbol. For example, we write $x > 7$ as $x \in (7, +\infty)$. The '(7' part indicates that no numbers equal to or lower than 7 are in the interval. The '$+\infty)$' part indicates that the interval continues forever in the positive direction. That is, there is no upper bound. So, the interval $(7, +\infty)$ is all numbers greater than 7. Similarly, the statement $w \in (-\infty, -2]$ is the same as $w \le -2$. Notice that we always use '(' with $-\infty$ and ')' with $+\infty$, instead of '[' and ']'.

$\boxed{9}$ Find all x such that $3x - 7 \ge 8 - 2x$.

10 Solve each of the following inequalities for respective variables.

(a) $2x + 15 \leq 21$

(b) $11 - 2t < 19 + 6t$

11 Solve the conjunctive inequality.

$$2 + x > 5 - 3x > -4$$

12 For what real values of x is the expression $3\sqrt{x} - 2$ between 10 and 16?

13 Solve $\dfrac{1}{2} < \dfrac{x}{x+1} < \dfrac{99}{100}$ for positive real number x.

14 Find the largest integer n satisfying

$$\frac{1}{2} < \frac{1}{2} \times \frac{2}{1} \times \frac{2}{3} \times \frac{3}{2} \times \cdots \frac{n}{n+1} < \frac{9}{10}$$

6.4 Graphing Linear Inequalities

Once we step up from one-variable to two-variable linear inequalities, it becomes impossible to write solutions with simple interval notation. We can, however, still graph the solutions. First, we draw the boundary line. Second, we plot any point that is not on the boundary. Third, color the proper region according to the result of the second step.

15 Graph the solution to the inequality $x - 2y \leq 4$.

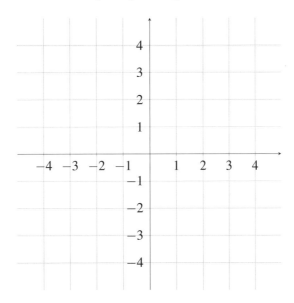

16 Shade the region of points that satisfy the inequality $x + 2y > -2$.

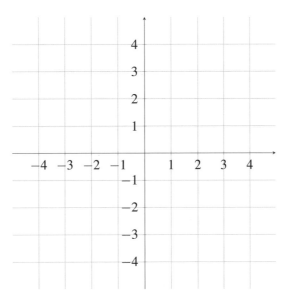

Determine whether a point $(2,3)$ is in the region shaded by $x - 2y > 3$.

Solution

If $(2,3)$ is in the region shaded by $x - 2y > 3$, then the inequality must hold as true. Let's check whether it works. Plugging $(x,y) = (2,3)$ into the inequality, we get $(2) - 2(3) = 2 - 6 = -4 < 3$. Hence, the inequality fails. This means that $(2,3)$ is not in the region shaded by $x - 2y > 3$.

17 Shade the region of points satisfying the following four conditions.

- $2x - 3y \leq 6$

- $x + 2y \geq 4$

- $x \geq 0$

- $y \geq 0$

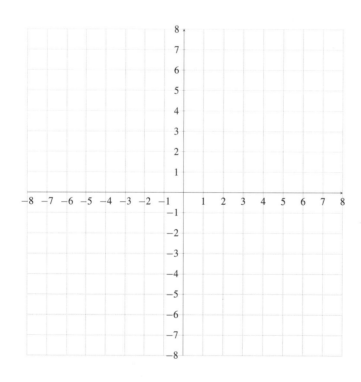

6.5 Linear Optimization

Optimization problems involve finding the minimum or maximum possible value of a quantity given certain constraints. Many optimization problems are naturally inequality problems, while others call on our logic skills or the clever usage of other mathematical tools.

Suppose we must minimize or maximize a linear expression in one or two variables such that the variables satisfy a set of linear inequalities. Suppose further that

- The set of all points that satisfy the inequalities is a bounded region.

- The inequalities are less-strict so that boundary points satisfy them.

Then, the point inside the bounded region that optimizes the desired expression must be a "corner," or vertex, of the region. Look at the following example.

18 Maximize $3x + y$ if $4x + 3y \leq 24$, $5x - 2y \geq -10$ and $y \geq -3$.

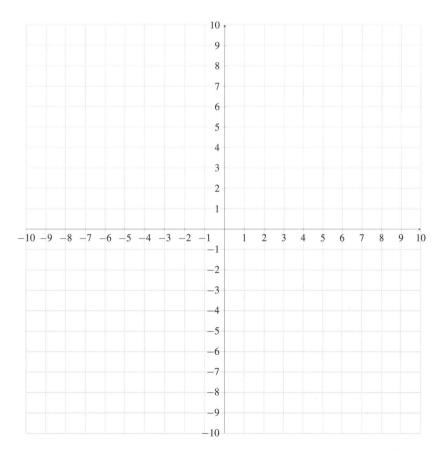

1 Graph the solution of the following inequalities on the number line and write the solution to the inequality using the interval notation.

(a) $2 + 3x \geq 14$.

(b) $2 + 5x < 15 - 8x$.

(c) $10 - 3x \leq 6 + 4x < 25 - 3x$.

2 Without using a calculator, compare 5^4, 6^3 and 7^2.

3 Suppose Bob the bibliophile[1] has a budget of $230, and he loves buying books. Assume he will buy as many books as he can, and the books Bob will buy each cost $17. At most, how many books can Bob afford and still have at least $25 left?

4 What is the smallest positive integer that has a square root that is greater than 10?

5 Find the maximum possible value for $x + y$, given that

- $3x + 2y \leq 6$

- $2x + 4y \leq 8$

- $x \geq 0$

- $y \geq 0$

☐1

(a) $[4, \infty)$

(b) $(-\infty, 1)$

(c) $\dfrac{4}{7} \leq x < \dfrac{19}{7}$

☐2 $7^2 < 6^3 < 5^4$

☐3 12

☐4 The smallest value of n is 101.

☐5 Use the vertex of the shaded region to find out $x + y$ is maximized at $(1, \dfrac{3}{2})$, so the maximum value is $\dfrac{5}{2}$.

Topic 7

Introduction to Quadratics

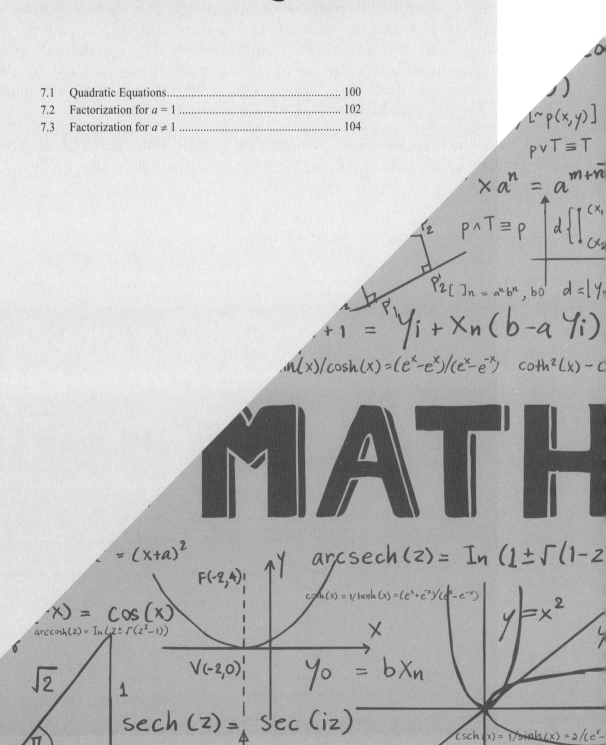

7.1 Quadratic Equations

In this chapter, we introduce another term and deal with equations of the form

$$ax^2 + bx + c = 0,$$

where a, b, and c are constants and $a \neq 0$. These equations are called quadratic equations, where the ax^2 term is called the quadratic term, the bx term is the linear term, and the c term is the constant term. The expression $ax^2 + bx + c$ is a quadratic expression, and a, b, and c are coefficients of the expression. We will often refer to a quadratic equation or expression simply as a quadratic. We also sometimes call a quadratic in which x is the variable a "quadratic in x." Similarly, $y^2 + 3y + 2$ is a quadratic in y and $3r^2 + 2r - 5$ is a quadratic in r.

We solved linear equations by isolating the variable. Unfortunately, having both an x^2 and an x term makes isolating the variable in a quadratic equation a little trickier.

$\boxed{1}$ Find the values of x that satisfy $x^2 = 16$.

$\boxed{2}$ Find the values of x that satisfy $2x^2 - 50 = 0$.

Let's practice expanding the product of two linear terms.

3 For each of the following parts, find the values of x for which the expression equals 0, and expand the product given.

(a) $(x-3)(x+5)$

(b) $(x-4)(-x+4)$

(c) $\left(x-\dfrac{2}{3}\right)\left(x+\dfrac{4}{3}\right)$

(d) $(x-2)(x+2)$

4 If $10000x = \dfrac{4}{x}$, where $x > 0$, what is the value of x?

7.2 Factorization for $a = 1$

We saw that if we can write a quadratic as the product of binomials, we can find the values of the variable for which the quadratic equals 0. We call this process of writing a quadratic as the product of binomial factors of the quadratic. We call each binomial in the product as a factor of the quadratic.

Example

Find all solutions to the equation $x^2 + 3x + 2 = 0$.

Solution
Let r and s be the solutions to the equation $x^2 + 3x + 2 = 0$. Then,

$$(x+r)(x+s) = x^2 + (r+s)x + rs$$
$$= x^2 + 3x + 2$$
$$= (x+1)(x+2)$$

Hence, $(r,s) = (1,2)$ or $(2,1)$. Either way, the solutions are $x = -1$ or $x = -2$.

$\boxed{5}$ Look at the parity(even or oddness) of the coefficient of x to solve $x^2 + 8x + 15 = 0$ for x.

$\boxed{6}$ Look at the parity of the coefficient of x to solve $x^2 - 5x - 24 = 0$ for x.

Example

Find all solutions to the equation $x^2 - 9x + 8 = 0$.

Solution

Let r and s be the solutions to the equation $x^2 - 9x + 8 = 0$. Then,

$$\begin{aligned}
(x+r)(x+s) &= x^2 + (r+s)x + rs \\
&= x^2 - 9x + 8 \\
&= (x-1)(x-8)
\end{aligned}$$

Hence, $(r,s) = (1,8)$ or $(8,1)$. Either way, the solutions are $x = 1$ or $x = 8$.

7 Find all solutions to each of the following equations.

(a) $x^2 - 11x + 28 = 0$

(b) $x^2 - 10x + 25 = 0$

(c) $x^2 - x - 56 = 0$

(d) $x^2 + 8x = 0$

7.3 Factorization for $a \neq 1$

Example

Find all solutions to the equation $2x^2 - 3x + 1 = 0$.

Solution

$$2x^2 - 3x + 1 = (2x - 1)(x - 1)$$

Hence, $x = \dfrac{1}{2}$ and $x = 1$.

8 Let a, b and c be numbers such that

$$(3x + 1)(x + 6) = ax^2 + bx + c$$

for all values of x. Find a, b and c. Hence, find all solutions to the equation $3x^2 + 19x + 6 = 0$.

9 Suppose $5x^2 + 18x - 35 = (ax + b)(cx + d)$ for all values of x.

(a) Find ac and bd.

(b) Find $ad + bc$.

(c) Hence, find the solutions to the equation $5x^2 + 18x - 35 = 0$.

10 If $(ax + b)(cx + d) = 12x^2 + 8x - 15$,

(a) find ac, bd, and $ad + bc$.

(b) Hence, solve $12x^2 + 8x - 15 = 0$.

11 Determine whether the coefficient of linear term of the following product is even or odd.

$$(2x-1)(2x+2)$$
$$(2x-17)(2x+3)$$
$$(4x+1)(x+3)$$

12 Solve the following questions.

(a) Bob, who learned about factoring quadratic expressions, figured that $8x^2 - 23x + 15 \neq (4x-3)(2x-5)$ by looking at the linear term. Explain how he knew about it without expanding the right side of the equation.

(b) Hence, solve $12x^2 + 28x - 5 = 0$.

13 Find all solutions to the equation

$$(3x - 2)(2x - 5) + (x - 7)(3x - 2) = 0.$$

14 For what positive values of k can $5x^2 + kx + 2$ be factored as the product of two linear terms with integer coefficients?

Given a quadratic equation $ax^2 + bx + c$ where r and s are solutions, then

$$ax^2 + bx + c = a(x - r)(x - s)$$

such that

$$r + s = -\frac{b}{a}$$

$$rs = \frac{c}{a}$$

Example

If $x^2 - Ax + 3 = 0$, where all roots are positive integers, what must be the value of A?

Solution

$$x^2 - Ax + 3 = (x - 3)(x - 1)$$
$$= (x - 1)(x - 3)$$

Hence, $x = 1$ and $x = 3$. Therefore, $A = 1 + 3 = 4$.

15 Consider the equation $x^2 - 7x + 12 = 0$.

(a) Find all solutions to the equation.

(b) Find the sum of the solutions and the product of the solutions.

16 The values $x = -2$ and $x = 5$ satisfy the equation $x^2 + Ax + B = 0$ where A and B are constants. What are the values of A and B?

17 Let r and s be the solutions of $19x^2 + 2x + 38 = 0$. Find the value of $(r-1)(s-1)$.

18 Let r and s be the roots of the equation $x^2 - 4x + 2 = 0$. If $r + \dfrac{1}{s}$ and $s + \dfrac{1}{r}$ are solutions to $x^2 - mx + n = 0$, what is the value of n?

1 Find all solutions to each of the following equations.

(a) $x^2 - 7x = 0$

(b) $x^2 + 3x = 7x - x^2$

(c) $2x^2 = 242$

(d) $16 - y^2 = -4$

2 Find all values of x satisfying $\dfrac{4}{x} = \dfrac{x}{16}$.

3 Find all solutions to the following equations.

(a) $x^2 - 8x + 7 = 0$

(b) $x^2 - 6x - 72 = 0$

(c) $4x^2 - 2x - 1 = 0$

(d) $x^2 - 13x + 12 = 0$

4 Find the values of the constants a and b such that $x = 4$ and $x = 3$ are both solutions to the equation $ax^2 + bx + 6 = 0$.

5 The equation $ax^2 + 5x - 3 = 0$ has $x = 1$ as a solution. What is the value of a? Hence, find the other solution.

6 Given a quadratic equation $35x^2 - 8x - 60 = 0$, solve the following questions.

(a) Find the sum and the product of the roots.

(b) Find the roots of the equation.

☐1

(a) $x^2 - 7x = x(x - 7) = 0$, so $x = 0$ or $x = 7$.

(b) $2x^2 - 4x = 2x(x - 2) = 0$, so $x = 0$ or 2.

(c) $2x^2 - 242 = 2(x^2 - 121) = 2(x - 11)(x + 11) = 0$, so $x = 11$ or $x = -11$.

(d) $20 = y^2$, so $y = \pm 2\sqrt{5}$

☐2 $x = \pm 8$

☐3

(a) $x = 7$ or $x = 1$

(b) $x = -6$ or $x = 12$

(c) Irreducible. This does not mean that there is no solution. In fact, solutions are $x = \dfrac{2 \pm \sqrt{4 + 16}}{4} = \dfrac{1 \pm \sqrt{5}}{2}$. Refer to Topic 10.3 for more details.

(d) $x = 12$ or $x = 1$

☐4 $a = \dfrac{1}{2}$ and $b = -\dfrac{7}{2}$

☐5 $a = -2$. Hence, the other solution is $x = \dfrac{3}{2}$.

☐6

(a) The sum of roots is $\dfrac{8}{35}$ and the product of roots is $-\dfrac{12}{7}$

(b) $x = -\dfrac{6}{5}$ or $x = \dfrac{10}{7}$.

Topic 8

More Factorization

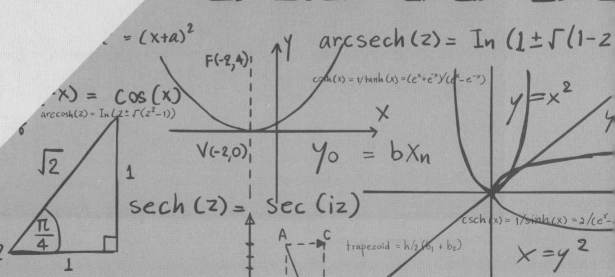

8.1 Square of Binomials

In the last chapter, we learned how to multiply binomials to form a quadratic and how to reverse the process. Now, we will look at special cases.

$$(\square \pm \triangle)^2 = \square^2 \pm 2 \cdot \square \cdot \triangle + \triangle^2$$

1 Expand each of the following square of binomials.

(a) $(x+5)^2$

(b) $(2x+3)^2$

(c) $\left(\dfrac{x}{2}-3\right)^2$

(d) $(5-2x)^2$

When we look at quadratics in perfect square form, always try to find the correct value that fills the box in the below expressions.

$$x^2 \pm 2 \cdot \square \cdot x + \square^2 = (x \pm \square)^2$$

2 For each of the following quadratics, determine whether the expression is a square of binomial. If so, factorize the expression.

(a) $x^2 - 8x + 8$

(b) $z^2 - 14z + 49$

(c) $x^2 - x + \dfrac{1}{4}$

(d) $4x^2 - 28x + 49$

3 If $x = \dfrac{\sqrt{6}}{3}$, evaluate $\left(x - \dfrac{1}{x}\right)^2$.

4 Compute the following expression without using a calculator.

$$20192020^2 - 2(20192020)(20192021) + 20192021^2$$

5 Square the following numbers without a calculator.

(a) $61^2 (= (60+1)^2)$ (b) $299^2 (= (300-1)^2)$

The triangle below is called Pascal's Triangle. After the 1 at the top, we generate each number in the triangle by adding the two numbers directly above it, connected by the line segments. Also, there are 1's in the first and the last spot of each layer.

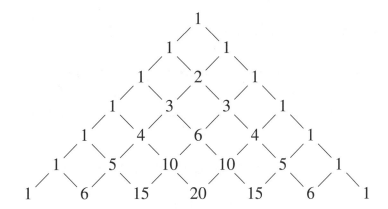

6 Expand the product $(x+y)^n$ for $n = 3$, 4, and 5. Find the patterns of the coefficients.

(a) $(x+y)^3$

(b) $(x+y)^4$

(c) $(x+y)^5$

8.2 Difference of Squares

Given $A^2 - B^2$, we get

$$(A - B)(A + B) = A^2 - B^2$$

The difference of squares is useful when we factor the expression, quite frequently appearing in contest questions, and it expedites the process of arithmetic. Before using the difference of squares, make sure that the terms are factored by grouping the common expressions between the two.

7 Factor the following expressions using the difference of squares.

(a) $x^2 - 4$

(b) $2x^2 - 8$

(c) $25x^2 - 16y^2$

(d) $144x^2 - 196y^2$

8 Evaluate the following product using the difference of squares.

(a) $6^2 - 5^2$

(b) $7^2 - 6^2$

(c) $11^2 - 10^2$

(d) $2020^2 - 2019^2$

9 Compute $\sqrt{2022 \cdot 2020 \cdot 2018 \cdot 2016 + 16}$ without a calculator.

8.3 Four-Term Quadratics

We can factorize the sum into the product of binomials, i.e.,

$$ab + ma + nb + mn = (a+n)(b+m)$$

which is also know as Simon's Favorite Factoring Technique, named after Art of Problem Solving user called Simon.

10 Find all pairs of positive integers m and n such that $mn + m + n = 76$.

11 Find all pairs of integers b and c that satisfy the equation $bc - 7b + 3c = 70$.

12 How many ordered pairs (m, n) of positive integers are solutions to

$$\frac{4}{m} + \frac{2}{n} = 1?$$

13 Find all pairs of integers (p, q) such that $pq - 3p + 5q = 0$.

1 Factor the following expressions using the squares of binomials.

(a) $4x^2 + 16x + 16$ (b) $x^2 - 10x + 25$

(c) $-x^2 - 6x - 9$ (d) $2x^2 + 28x + 98$

(e) $9x^2 - 6x + 1$ (f) $x^2 + \dfrac{7}{4}x + \dfrac{49}{64}$

2 Factor each of the following binomials, using the difference of squares.

(a) $x^2 - 16$

(b) $225 - 4y^2$

(c) $20x^2 - 5y^2$

(d) $16x^2 - 9y^2$

3 Solve the following quadratic equations, using the difference of squares.

(a) $36x^2 = 16$

(b) $\dfrac{1}{4}x^2 = \dfrac{1}{25}$

4 If b is a constant such that $x^2 + bx + 100$ is the square of binomial, what could be the two values of b?

5 Which consecutive perfect squares have a difference of 63?

6 Use SFFT to factor the following sum.

(a) $xy - 7x + 9y - 63$

(b) $-xy + 2x + 5y - 10$

7 Find all pairs of integers (x, y) that satisfy the equation $xy - 2x + 7y = 49$.

1

(a) $4(x+2)^2$

(b) $(x-5)^2$

(c) $-(x+3)^2$

(d) $2(x+7)^2$

(e) $(3x-1)^2$

(f) $(x+\dfrac{7}{8})^2$

2

(a) $(x-4)(x+4)$

(b) $(15-2y)(15+2y)$

(c) $5(2x-y)(2x+y)$

(d) $(4x-3y)(4x+3y)$

3

(a) $x = \pm\dfrac{2}{3}$

(b) $x = \pm\dfrac{2}{5}$

4 $b = \pm 20$

5 $32^2 - 31^2 = 63$, so the larger is 32 and the smaller is 31.

6

(a) $(x+9)(y-7)$

(b) $-(x-5)(y-2)$

7 $(x,y) = (-6,37), (-2,9), (0,7), (28,3), (-8,-33), (-12,-5), (-14,-3), (-42,1)$

Topic 9

Complex Number

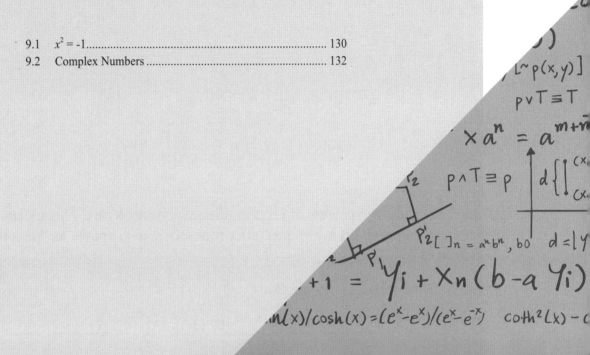

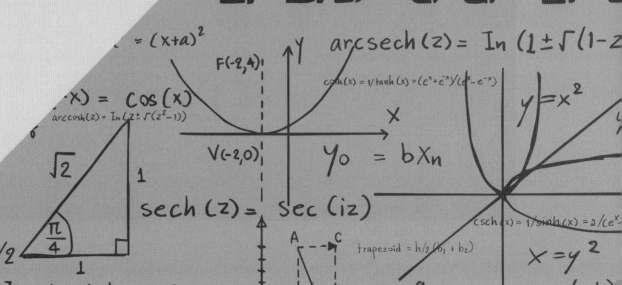

A square of real number cannot be negative. However, what if it is not real? An imaginary number is a number whose square is negative. We use the symbol i to denote the number whose square is -1, i.e., $i^2 = -1$. The numbers we have used prior to this chapter in this text are real numbers that can be plotted on a real number line for comparison. In this topic, we shall expand the set of real numbers into the set of complex numbers that allow negative number inside the radicand. Let's investigate powers of i, and its peculiar property appears at the fourth power. Have a look at it.

$$i = i$$
$$i^2 = -1$$
$$i^3 = -i$$
$$i^4 = 1$$
$$\vdots$$

As one can check, the powers of i repeat after every four terms. This means that power expressions involving i repeat after four terms, also known as "periodic with the period of 4."

1 Evaluate the following expressions.

(a) $(-2i)^2$

(b) $(3i)^2$

(c) i^{100}

(d) i^{-20}

We can use imaginary numbers to factorize a quadratic expression that is irreducible in real numbers.

Example

Find all complex solutions to the following equation

$$x^2 + 4 = 0$$

Solution
$x^2 = -4$ implies $x = \pm\sqrt{-4} = \pm 2i$. There are two complex solutions $x = 2i$ or $x = -2i$.

2 Find all complex solutions to each of the following equations.

(a) $x^2 + 25 = 0$ 　　　　　　　　　　　(b) $x^2 + 15 = 0$

3 Simplify the following powers of i.

(a) i^{-124} 　　　　　　　　　　　(b) i^{2019}

9.2 Complex Numbers

$$z = x + yi$$

Any number of the form $x + yi$, where x and y are real numbers, is called a complex number, usually denoted by z. The real number x is the real part of the complex number $x + yi$, known as $\text{Re}(z)$, and the real number y is the imaginary part of the complex number, known as $\text{Im}(z)$.

We can add two complex numbers by adding the real parts of the two complex numbers and adding the imaginary parts. For example,

$$(1 + 3i) + (-7 - 4i) = (1 - 7) + (3i - 4i) = -6 - i.$$

We can multiply two complex numbers using the distributive property of multiplication. For example,

$$\begin{aligned}
(1 + 3i)(-7 - 4i) &= 1(-7 - 4i) + 3i(-7 - 4i) \\
&= (1)(-7) + (1)(-4i) + (3i)(-7) + (3i)(-4i) \\
&= 5 - 25i.
\end{aligned}$$

Also, a complex number whose imaginary part is 0 is called a real number. On the other hand, a complex number whose real part is 0 is called a purely imaginary number. If both real part and imaginary part are 0, then we simply call it 0.

4 Simplify each of the following.

(a) $(5 + 3i) + (2 - 3i)$

(b) $(3 - 2i) + (3 + 2i)$

(c) $i + (2 + 3i) + (4 - 4i)$

(d) $(2 - 3i) + (4 - i) + (-6 + 4i)$

5 Expand each of the following products.

(a) $(2-i)(3-2i)$ (b) $(4-i)(4+i)$

6 Suppose A is a real number, and the product $(2+5i)(2+Ai)$ is also a real number. Find A and the value of this product.

$$\bar{z} = x - yi$$

The number that results when the sign of the imaginary part of a complex number is reversed is called the complex conjugate of the original number. We usually refer to a complex conjugate as just a conjugate. For example, $7 + 3i$ is the conjugate of $7 - 3i$ and $-3 - 8i$ is the conjugate of $-3 + 8i$. If z is a complex number, then we denote the conjugate of z as $\bar{z}$. For example,

$$\overline{1 + 2i} = 1 - 2i.$$

We use the conjugates when we rationalize the denominator. In other words, if there is a complex number $a + bi$ in the denominator, make sure we multiply $a - bi$ to both numerator and denominator to rationalize the denominator.

7 Write the quotient $\dfrac{1}{2 - i}$ as a single complex number.

8 Write the quotient $\dfrac{1 + i}{3 - i}$ as a single complex number.

1 Compute the following complex expressions.

(a) $(2i)^2$

(b) $(-3i)^3$

2 Compute the following sum of powers of i.

$$i^{2019} + i^{2018} + \cdots + i + 1$$

3 Simplify the following expressions.

(a) $(5+2i)-(-3-5i)$

(b) $4(2+7i)+3(4-i)$

4 Simplify the following expressions.

(a) $(5-3i)(-4+3i)$

(b) $(-1+5i)(2+8i)(1+5i)$

5 If A is real and the product of $(A+6i)(7-3i)$ is a real number, then find the value of A.

6 The number $14+i$ can be factored into the product of $3-2i$ and what other complex number?

7 Simplify the following complex expression.

$$(i-i^{-1})^{-2}$$

1

(a) -4

(b) $27i$

2 0

3

(a) $8 + 7i$

(b) $20 + 25i$

4

(a) $-11 + 27i$

(b) $-52 - 208i$

5 $A = 14$

6 $\dfrac{40 + 31i}{13}$

7 $-\dfrac{1}{4}$

Topic 10

Completing the Squares

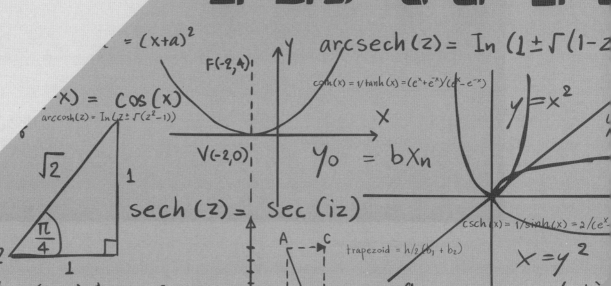

10.1 The Perfect Squares

We investigated squares of binomials and discovered the relationship

$$a^2 + \underline{2b}a + \underline{b}^2 = (a+b)^2$$

This relationship can be used to solve any quadratic equation. Over the next two sections, we'll learn how. Solving a quadratic equation is simple when one side of the equation involves a perfect square. For instance,

$$A^2 = B$$
$$\sqrt{A^2} = \sqrt{B}$$
$$|A| = \sqrt{B}$$
$$A = \pm\sqrt{B}$$

As long as we have the form of $A^2 = B$, we can always find the solution relevant to the given quadratic equation.

1 Find all complex solutions to the following equations.

(a) $x^2 - 16 = 0$ (b) $(x+2)^2 - 16 = 0$

(c) $x^2 - 12 = 0$ (d) $(x+2)^2 + 9 = 0$

$\boxed{2}$ Given a quadratic equation $x^2 + 2x - 7 = 0$, find all of the solutions.

$\boxed{3}$ If the quadratic expression $x^2 + 32x + c$ can be written as the square of a binomial, $(x+a)^2$, for some constants a and c, what is the value of a?

4 Find the values of the constants a and c for the following parts.

(a) $x^2 - 6x + c$ equals $(x+a)^2$

(b) $r^2 + 3r + c$ equals $(r+a)^2$

(c) $x^2 - \dfrac{x}{2} + c$ equals $(x+a)^2$

10.2 Completing the Squares

We know how to solve equations of the form $(x+a)^2 + b = 0$, i.e., $A^2 = B$ form. Furthermore, we know how to figure out what constant term is needed to make a quadratic a perfect square. In this section, we learn how to put these two skills together to solve any quadratic equation of x.

Example

Complete the square for $x^2 + 4x - 1 = 0$.

Solution

$$\begin{aligned} x^2 + 4x - 1 &= (x^2 + 4x + 4 - 4) - 1 \\ &= (x^2 + 4x + 4) - 5 \\ &= (x+2)^2 - 5 \end{aligned}$$

5 If $x^2 + 8x = 14$, complete the square to find the values of x satisfying the given equation.

6 If $3(x^2) + 3(4x) + 1 = 0$, find the solutions to the equation by completing the square.

Complete the square for $\underline{2}x^2 + \underline{4}x - 1 = 0$ by using the underlined parts to find out the constant that we add and subtract within the expression.

Solution

$$
\begin{aligned}
2x^2 + 4x - 1 &= 2(x^2 + 2x) - 1 \\
&= 2(x^2 + 2x + 1 - 1) - 1 \\
&= 2(x^2 + 2x + 1) - 2 - 1 \\
&= 2(x + 1)^2 - 3
\end{aligned}
$$

7 Solve each of the following equations by completing the square.

(a) $x^2 + 2x + 13 = 0$

(b) $12x^2 - 11x - 36 = 0$

10.3 Quadratic Formula

The general(or standard) form of a quadratic expression written in x is $ax^2 + bx + c$, where a, b, and c are constants and $a \neq 0$. In completing the square, we have a process of finding the proper value we add and subtract within the expression to turn it into the perfect square form. Abstracting this idea, we should be able to use completing the square to develop a formula for the solutions to a quadratic equation in terms of the coefficients, a, b, and c, of the quadratic. Given $ax^2 + bx + c = 0$, the quadratic formula tells us that

$$x = \frac{-b \pm \sqrt{b^2 - 4ac}}{2a}$$

- $b^2 - 4ac > 0$: two real solutions

- $b^2 - 4ac = 0$: one real solution

- $b^2 - 4ac < 0$: no real solution

$\boxed{8}$ Let a, b, and c be constants, where $a \neq 0$. Deduce the quadratic formula for $ax^2 + bx + c = 0$.

9 Find all complex solutions of each of the following quadratic equations by using the quadratic formula.

(a) $x^2 + 4x - 96 = 0$

(b) $9x^2 - 42x + 49 = 0$

(c) $3x^2 - 14x + 8 = 0$

(d) $x^2 + 6ix - 5 = 0$

Given a quadratic equation $ax^2 + bx + c = 0$, the expression $b^2 - 4ac$ determines whether the equation has a real solution or not. If it is positive, then the equation has two distinct real solutions. If it is 0, then the equation has one real solution. Lastly, if it is negative, then the equation has no real solution.

10 Bob says that

$$x^2 + x = -11.$$

has two real solutions. Use the quadratic formula to find a way to determine if the roots of a quadratic with real coefficients are real or not. Can you use your method to determine whether Bob is correct?[1]

11 The quadratic equation $ax^2 - 5x + 6 = 0$ has a double(repeated) root for some value of a. What is that root?

---

[1]This is called the discriminant of quadratics.

If r and s are solutions to the equation $ax^2 + bx + c = 0$, then the following three properties are automatically satisfied.

- $r = \dfrac{-b + \sqrt{b^2 - 4ac}}{2a}$, and $s = \dfrac{-b - \sqrt{b^2 - 4ac}}{2a}$

- $ar^2 + br + c = as^2 + bs + c = 0$

- $ax^2 + bx + c = a(x - r)(x - s)$ so $r + s = -\dfrac{b}{a}$ and $rs = \dfrac{c}{a}$

Example

Find a quadratic equation that has the solution of $x = 1 - i$

Solution

$$x = 1 - i$$
$$x - 1 = -i$$
$$(x - 1)^2 = (-i)^2$$
$$x^2 - 2x + 1 = -1$$
$$x^2 - 2x + 2 = 0$$

12 If $x = 1 - 3i$ is a solution to $x^2 + bx + c = 0$, manipulate the linear equation to retrieve the values of b and c.

13 If the roots, s and t, of $2x^2 - mx - 8 = 0$ differ by $m - 1$, find all possible values of m.

14 Let r and s be the roots of the quadratic $x^2 + bx + c$, where b and c are constants. If $(r - 1)(s - 1) = 7$, find $b + c$.

1 Solve each of the following equations for x.

(a) $(x-5)^2 - 27 = 0$

(b) $2(5-3x)^2 - 12 = 0$

2 In each part below, find the positive constant A that makes the quadratic the square of a binomial.

(a) $2x^2 + Ax + 242$

(b) $x^2 + \dfrac{10}{3}x + A$

3 Complete the square to solve the following equations.

(a) $x^2 - 5x - 3 = 0$

(b) $10x^2 + 6x - 5 = 0$

4 Suppose that $(m-1)x^2 + mx + 8 = 0$ has a double root (or repeated root). Find all possible values of m satisfying this double-root condition.

5 Find all values of x that satisfy $2x + \dfrac{3}{2-x} + 5 = 0$.

6 Find all values of x such that $\dfrac{x}{x-1} - \dfrac{2}{x-3} = 1$.

7 Use the quadratic formula to find the complex roots of $x^2 + ix - 2 = 0$.

8 The product of the roots of the quadratic equation $5x^2 + Ax - 2 = 0$ is 4 less than the sum of the roots, and A is a constant. What is the value of A?

1

(a) $x = 5 \pm 3\sqrt{3}$

(b) $x = \dfrac{5 \pm \sqrt{6}}{3}$

2

(a) $A = 44$

(b) $A = \dfrac{25}{9}$

3

(a) $x = \dfrac{5 \pm \sqrt{37}}{2}$

(b) $x = \dfrac{-3 \pm \sqrt{59}}{10}$

4 $m = 16 \pm 4\sqrt{14}$

5 $x = \dfrac{-1 \pm \sqrt{105}}{4}$

6 $x = -1$

7 $x = \dfrac{-i \pm \sqrt{7}}{2}$

8 $A = -18$

Topic 11

Graphing Quadratics

and Inequalities

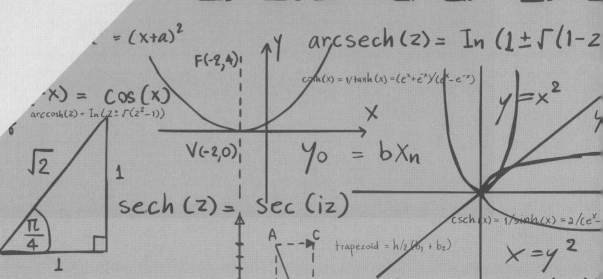

11.1 Parabola

We will learn how to graph the graphs of quadratic equations. We start from plotting the graph of $y = x^2$. The shape of the graph of quadratic equations is called a parabola.

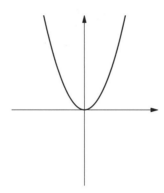

- The y-axis is the axis of symmetry.

- $(0,0)$ is the vertex of the parabola.

1 Graph the equations $y = \dfrac{1}{2}x^2$ and $y = -2x^2$.

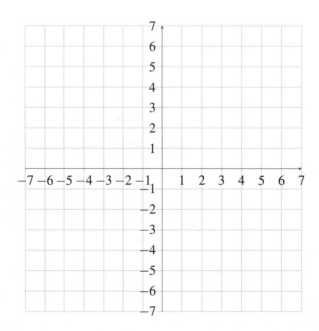

Given a quadratic function $y = ax^2 + bx + c$, it can be transformed into the form called "vertex form", i.e.,

$$y = a(x-h)^2 + k$$

where

- $x = h$ is the axis of symmetry.

- If $a > 0$, then k is the minimum value. If $a < 0$, then k is the maximum value.

Example

Find the vertex form of $y = 2x^2 - 4x + 3$, and determine the axis of symmetry.

Solution

$$\begin{aligned}
y &= 2x^2 - 4x + 3 \\
&= 2(x^2 - 2x) + 3 \\
&= 2(x^2 - 2x + 1 - 1) + 3 \\
&= 2(x^2 - 2x + 1) - 2 + 3 \\
&= 2(x-1)^2 + 1
\end{aligned}$$

Hence, the x-coordinate of the vertex is the axis of symmetry. Hence, $x = 1$.

$\boxed{2}$ Graph $y = \dfrac{1}{2}(x-2)^2 + 3$.

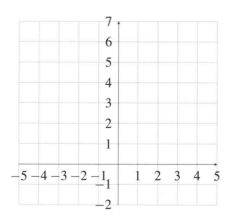

What if we want to graph $y = ax^2 + bx + c$? Let's start from the case when $a = 1$. The process of graphing it starts by plotting a few points. We will see the general shape of the graph, but not exactly where the vertex is. However, plotted points will tell us where the axis of symmetry is. Then, we can easily find the y-coordinate of the vertex.

For instance, given $y = x^2 - 2x + 2$. We can produce the following table values of x and y.

x	-2	-1	0	1	2	3
y	10	5	2	1	2	5

If we plot the points, we notice that $x = 1$ is the axis of symmetry. Hence, the y-coordinate of the vertex is given by 1. Therefore, the equation $y = x^2 - 2x + 2$ is, in fact, $y = (x-1)^2 + 1$. Let's practice graphing the curve using the same set of strategies.

$\boxed{3}$ Graph the equation $y = x^2 - 5x + 3$.

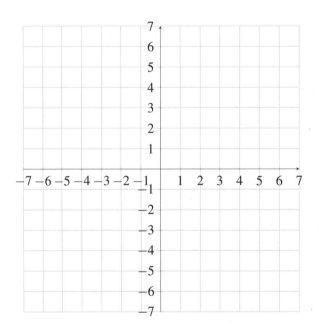

In fact, the transition from $y = ax^2 + bx + c$ into $y = a(x-h)^2 + k$ requires the process of completing the squares. Even if the coefficient of x^2 is not 1, we can simply complete the square to find the vertex and then graph it.

However, the shape of the graph may change as the coefficient of x^2 is positive or negative.

- If the coefficient of x^2 is positive, then the parabola opens upward.

- If the coefficient of x^2 is negative, then the parabola opens downward.

Example

Determine the concavity of $y = -x^2 + 4x - 3$, and locate the maximum(or minimum) point of the parabola.

Solution

$$y = -x^2 + 4x - 3$$
$$= -(x-2)^2 + 4 - 3$$
$$= -(x-2)^2 + 1$$

The graph is concave down, so the curve has its maximum point at the vertex $(2, 1)$.

4 Graph the equation $y = -2x^2 - 4x + 1$.

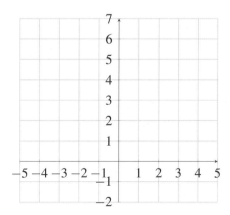

If x and y are switched, then the parabola $x = y^2$ opens to the left or right, instead of up or down. If the coefficient of y^2 is positive, then the parabola opens to the right. If it is negative, then the graph opens to the left.

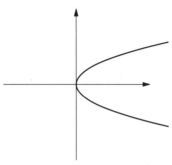

The graph of $x = y^2$

As we plot the graph of $x = y^2$, we continue plotting the points until there are two different values of y for the same value of x.

5 Graph the equation $x = y^2 + 4y + 1$.

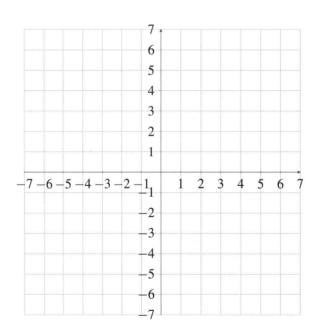

11.2 Circle

Circle is a set of points equidistant from a given point called the center. The common distance between points on the circle and the center of the circle is called the radius of the circle.

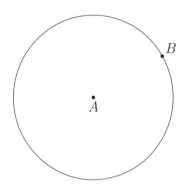

Given the center (h, k) and the radius r, the equation is given by

$$(x - h)^2 + (y - k)^2 = r^2$$

6 Find the equation whose graph is a circle with center $(3, 4)$ and radius 5.

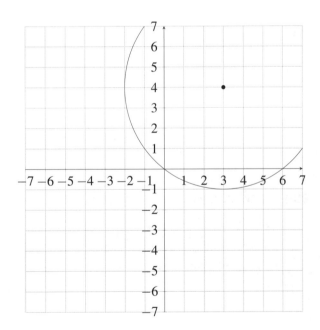

Find the center of $x^2 - 2x + y^2 + 2y = 0$

Solution

$$x^2 - 2x + y^2 + 2y = 0$$
$$(x^2 - 2x + 1) + (y^2 + 2y + 1) = 0 + 1 + 1$$
$$(x - 1)^2 + (y + 1)^2 = 2$$

The center is $(1, -1)$ with the radius of $\sqrt{2}$.

7 Given $(x - 2)^2 + (y - 3)^2 = 36$, find the center and the radius of the circle.

8 Find the center and the radius of the circle $x^2 - 4x + y^2 - 8y = 5$.

9 Find the center and radius of the circle satisfying

$$3x^2 + 6x + 3y^2 - 9y = -2$$

Example

Find the equation of the circle that passes through $(0,0)$, $(3,-1)$ and $(-1,7)$.

Solution

#1. Come up with the equation $(x-h)^2 + (y-k)^2 = r^2$, where (h,k) is the center and r is the radius.

#2. Substitute $(0,0)$, $(3,-1)$ and $(-1,7)$ into the equation.

#3. Solve $h^2 + k^2 = r^2$, $(3-h)^2 + (-1-k)^2 = r^2$, and $(-1-h)^2 + (7-k)^2 = r^2$.

#4. Eliminating $h^2 + k^2 = r^2$ from the last two equations, we get

$$-6h + 2k = -10$$
$$2h - 14k = -50$$

#5. Solving the system of equations above, we get $(h,k) = (3,4)$ and $r = 5$. Hence, the equation results in $(x-3)^2 + (y-4)^2 = 5^2$.

10 Find the circle equation that passes through three points $(0,0)$, $(3,-1)$ and $(-1,-7)$, where the answer may involve <u>fractions</u> and <u>radicals</u>.

11 Do the graphs of the equations $x - y = 8$ and $x^2 + y^2 - 4x + 12y + 28 = 0$ intersect? If so, at what points do the two graphs meet?

12 A diameter of a circle is a segment with endpoints on the circle such that the segment passes through the center of the circle. In the Cartesian plane, the segment with endpoints $(-6, 0)$ and $(16, 0)$ is the diameter of a circle. Find the x-coordinate of the center.

11.3 Quadratic Inequalities

If we have quadratic inequalities instead of equations, we can solve the inequality by

- sign analysis : case enumerating on the critical values found by factorization.

- graphing inequalities : comparing the portion of the parabola and the line to see if the portion is either above or below the given line.

Example

Solve $(x-1)(x-2) > 0$ by sign analysis.

Solution
#1. Critical values of $x = 1$ and $x = 2$.
#2. There are three possible cases to work on.
#3. If $x < 1$, then $(x-1)(x-2) > 0$. This must be included in the solution set.
#4. If $1 < x < 2$, then $(x-1)(x-2) < 0$. This interval is excluded from the solution set.
#5. If $2 < x$, then $(x-1)(x-2) > 0$. This is included in the solution set.
#6. Hence, the solution must be $x < 1$ or $2 < x$.

---

13 Solve the following inequality $(x+2)(x-1) \geq 0$ by sign analysis.

Example

Solve $(x-1)(x-2) > 0$ by graphing the graph of $y = (x-1)(x-2)$ in the plane.

Solution

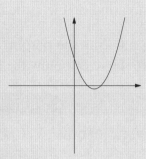

#1. We will compare the graph of $y = (x-1)(x-2)$ and $y = 0$.

#2. Determine which portion of the graph is above the x-axis.

#3. The solution must be when $x < 1$ or $2 < x$.

---

14 Find all values of x that satisfy $x^2 - 8x + 12 < 0$ by graphing the parabola in the plane.

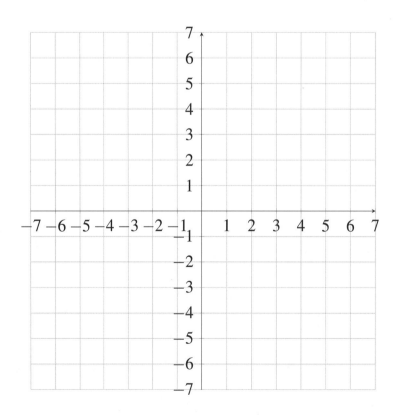

15 Graph the equation $y = x^2 - 6x + 13$.

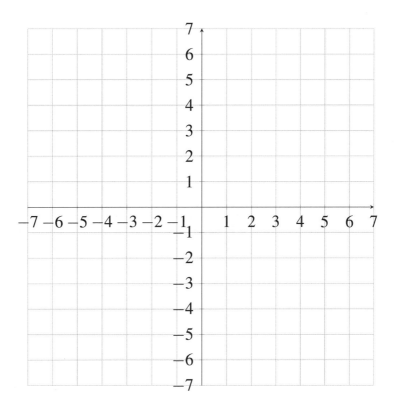

Explain why the inequality $x^2 - 6x + 13 < 0$ has no real solution.

16 Graph the inequality $y > x^2 - 4x - 2$.

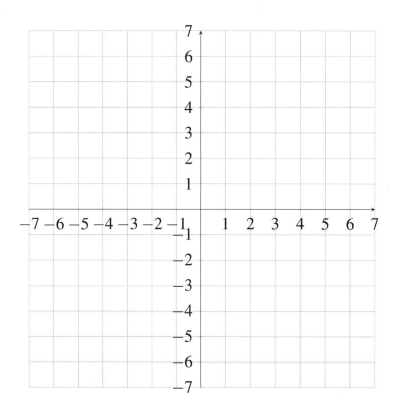

17 Find all values of k such that $x^2 - 6x + k \geq 5$ for all values of x.

11.4 Trivial Inequalities

If x is real, then
$$x^2 \geq 0$$
where $x^2 = 0$ if and only if $x = 0$. This inequality provides us with the tool of lower bounds. That being said, lower bounds or upper bounds can also be found by AM-GM inequality, given two real numbers. Next two examples will show us the definition and application of AM-GM inequality. First off, we have to know what AM and GM stand for.

- AM of x and y : arithmetic mean of x and y, i.e., $\dfrac{x+y}{2}$

- GM of x and y : geometric mean of x and y, i.e., $\sqrt{xy}$

18 Prove that $\dfrac{x^2+y^2}{2} \geq xy$ for any real x and y.

19 Prove that the sum of any positive real number and its reciprocal must be greater than or equal to 2.

11.5 Quadratic Optimization

Normally, when we solve for optimal values of quadratic expressions, we change the given expression into the vertex form. The optimal value of a quadratic expression is either maximum or minimum value, i.e., the y-coordinate of the vertex.

20 What is the largest possible value of $-x^2 + 6x - 7$, where x is a real number?

Sometimes, the answer may change if the domain is restricted so that it does not include the vertex. Then, the key idea to find out the maximum or minimum value is to look at the endpoints.

21 Find the smallest possible value of $2x^2 + 8x - 9$ if

(a) for all real x.

(b) for $0 \leq x \leq 4$.

22 Bob the football player kicks a ball off a platform that is 45 feet from the ground. The height in feet of the ball above the ground at t seconds after he kicks it is given by

$$h(t) = -16t^2 + 48t + 45$$

until the ball hits the ground, in feet.

(a) After how many seconds does the ball hit the ground?

(b) What is the greatest height the ball reaches?

23 Bob the agronomist(=farmer) has 20 meters of fencing to build a chicken run for his chickens. He can use the side of his barn as one side of the chicken run. What is the largest area his chicken run can be if it must be rectangular?

1 The parabola shown below is the graph of the equation $x = ay^2 + by + c$. Find the value of $a + b + c$.

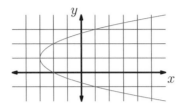

2 How many y-intercepts does the graph of the parabola $x = 2y^2 - 3y + 7$ have?

3 As shown in the figure below, the zeros of the quadratic $y = ax^2 + bx + c$ that passes through $(4, 12)$ with the vertex of $(2, -4)$ are located at $x = m$ and $x = n$, where $m > n$. What is the value of $m^2 + n^2$?

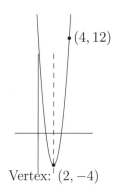

$(4, 12)$

Vertex: $(2, -4)$

4 For what value of k does the equation $x^2 + 10x + y^2 + 6y - k = 0$ represent a circle of radius 6?

5 The circle centered at $(2, -1)$ and with radius 4 intersects the circle centered at $(2, 5)$ and with radius $\sqrt{10}$ at two points A and B. Find $(AB)^2$. (Point A is not on the y-axis, although it looks like it.)

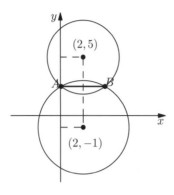

6 Suppose there is a circle whose equation is given by $x^2 - 6x + y^2 + 2y = 6$, which is inscribed inside a square. What is the area of the square?

7 Solve the inequality $x^2 - 4x - 21 \le 0$.

8 Find all values of x such that $x^2 - 8x + 16 > 0$.

9 Prove that $x^2 + 4 \geq 4x$ for all real numbers x.

10 If $(x - 2y)^2 + (y - 3)^2 = 0$, and x and y are real, then what is the value of xy?

$\boxed{1}$ -3

$\boxed{2}$ 0

$\boxed{3}$ 10

$\boxed{4}$ $k = 2$

$\boxed{5}$ $(AB)^2 = 15$

$\boxed{6}$ 64

$\boxed{7}$ $-3 \le x \le 7$

$\boxed{8}$ The set of all real numbers except $x = 4$

$\boxed{9}$ For all real x, $(x-2)^2 \ge 0$. Hence, $x^2 - 4x + 4 \ge 0$. Therefore, $x^2 + 4 \ge 4x$.

$\boxed{10}$ $xy = 18$

Topic 12

Functions

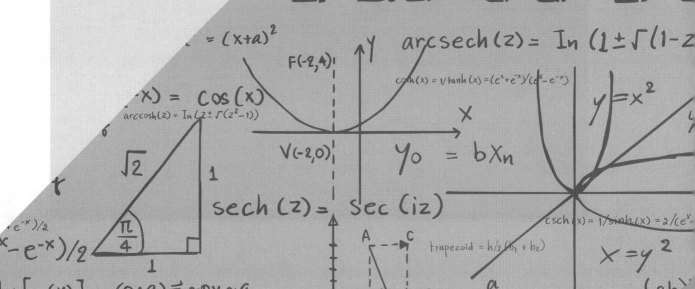

12.1 Function

Function is a machine that eats input and spits out output. It's quite trustworthy. It only spits out one output per one input. If this property is not satisfied, then we say it is NOT a function. We have the following illustration to understand the concept of "function."

$$f(\text{ input }) = \text{ output}$$

Normally, we write $f(x) = y$ or $f : x \to y$, indicating that x is an input and y is the output for that particular input. Nevertheless, a function f can eat anything other than x.

Example

Write $f(1-x)$ in terms of x, if $f(x) = 2 - x$.

Solution

$$\begin{aligned} f(1-x) &= 2 - (1-x) \\ &= 2 - 1 + x \\ &= 1 + x \end{aligned}$$

$\boxed{1}$ Consider a <u>linear function</u> of x given by $f(x) = 4x - 2$. Solve the following questions.

(a) Find $f(1)$, $f(2)$ and $f(3x)$.

(b) For what value of x does $f(x) = -2$?

(c) Find $f(2k - 3)$.

Now, there are some vocabularies associated to "function."

- Domain is the set of x-values, i.e., the set of inputs.

- Range is the set of y-values, i.e., the set of outputs.

Since we learned about the quadratic graphs, let's have a look at a quadratic function.

Example

Find the domain and range of $y = x^2$.

Solution
Since there is no restriction on x, the domain is $\mathbb{R}$. On the other hand, by trivial inequality, $x^2 \geq 0$ for all real values of x. Hence, $y = x^2 \geq 0$. Therefore, the range is the set of non-negative real numbers.

$\boxed{2}$ Consider a quadratic function of x given by $f(x) = x^2 + 3$. Solve the following questions.

(a) Find $f(1)$, $f(-1)$ and $f(-2)$.

(b) Is there a real value of x such that $f(x) = 4$? How about $f(x) = 2$?

(c) State the range and the domain of f.

There are functions other than linear or quadratic functions. Let's have a look at a rational function, whose parent function is $y = \dfrac{1}{x}$.

Find the domain and range of $y = \dfrac{1}{x}$.

Solution
Since $x \neq 0$, the domain must be the set of all real numbers except 0. Now, $y = \dfrac{1}{x} \implies x = \dfrac{1}{y}$. Hence, in order for x to exist as a real number, $y \neq 0$. Therefore, the range must be the same set as the domain.

$\boxed{3}$ Consider a <u>rational function</u> of x given by $f(x) = \dfrac{x-2}{2x+3}$. Solve the following questions.

(a) Find $f(1)$, $f(0)$, and $f\left(-\dfrac{1}{2}\right)$.

(b) Are there any values of x that do not produce a value of $f(x)$? What is the domain of f?

(c) Try to find a value of x such that $f(x) = \dfrac{1}{2}$. Now, suppose $y = \dfrac{x-2}{2x+3}$. Solve for x in terms of y. Use this to find out the range of f.

Function may have a restriction on the domain. In particular, we will have a look at linear functions when they have a restriction on the domain.

Example

Given a linear function $f(x) = x + 2$ for $1 \leq x \leq 3$, find its range.

Solution

$$1 \leq x \leq 3$$
$$1 + 2 \leq x + 2 \leq 3 + 2$$
$$3 \leq f(x) \leq 5$$

---

4 Suppose we have $h(x) = 2 - 3x$ for $2 \leq x \leq 9$. (In other words, we do not have $h(0)$, even though we are able to compute the value.)

(a) Find the value of $2 - 3x = 1$. Is it in the domain of h? Is 1 in the range of h?

(b) What is the range of h?

Function does not have to eat one input. In fact, it can eat multiple inputs at the same time, but the golden rule is that its output is always one value. Given a multivariable function, place values in proper placeholders.

Example

Compute $f(2,3)$ if $f(x,y) = 2x + 3xy$.

Solution
$f(2,3) = 2(2) + 3(2)(3) = 4 + 18 = 22.$

$\boxed{5}$ Suppose $g(x,y,z) = 2x - 3z + 4y$. Find y if $g(1,y,2) = 11$.

$\boxed{6}$ If $f(x,y) = x^2 + 2x + y^2 - 4y + 5$, what is the smallest possible value of $f(x,y)$?

12.2 Function Arithmetic

Given any two function, $f(x)$ and $g(x)$, we can add, subtract, multiply, or divide functions, producing a new function with new domain and range.

Example

Evaluate $f(2) \times g(2)$ where $f(x) = |x-2|$ and $g(x) = x^2 - 1$.

Solution

$$f(2) \times g(2) = |2-2| \times (2^2 - 1)$$
$$= 0 \times 3$$
$$= 0$$

7 If $f(x) = 2x + 3$ and $g(x) = 3 - 7x$.

(a) Compute $f(1) + g(1)$.

(b) Compute $f(2) - g(2)$.

(c) Compute $f(4) \cdot g(4)$.

(d) Compute $\dfrac{f(2)}{g(2)}$.

What if each function has different domains? We will have a look at the following example, step-by-step.

Example

Find the domain of $f(x) + g(x)$ where $f(x) = \sqrt{2-x}$ and $g(x) = \sqrt{x-1}$.

Solution
#1. The domain of $f(x)$ is given by $2 - x \geq 0$.
#2. The domain of $g(x)$ is given by $x - 1 \geq 0$.
#3. Hence, the intersection between the two intervals is the domain of $f(x) + g(x)$.
#4. The answer is the set of all real x such that $1 \leq x \leq 2$.

$\boxed{8}$ Let $f(x) = \sqrt{x}$ and $g(x) = \sqrt{x^2 - x - 6}$. Find

(a) the domain of $f(x)$;

(b) the domain of $g(x)$;

(c) the domain of $s(x) = f(x) + g(x)$;

(d) the domain of $d(x) = f(x)/g(x)$.

Example

Let $f(x) = \sqrt{x}$ and $g(x) = \sqrt{x-1}$. Let $h(x) = (f \cdot g)(x) = f(x) \cdot g(x)$. Since $\sqrt{x}\sqrt{x-1} = \sqrt{x^2-x}$, is $\sqrt{x^2-x}$ defined when $x = -1$? Is -1 is the domain of h?

Solution

#1. When $x = -1$, then $\sqrt{(-1)^2 - (-1)} = \sqrt{2}$, so it is indeed defined at $x = -1$.

#2. However, -1 is not in the domain of either f or g.

#3. Hence, the value -1 is not in the domain of h.

9 Let $f(x) = \sqrt{2-x}$ and $g(x) = \sqrt{x}$.

(a) Find the domain of $f(x) \pm g(x)$.

(b) Find the domain of $f(x) \cdot g(x)$.

(c) Find the domain of $\dfrac{f(x)}{g(x)}$.

12.3 Composition of Two Functions

In the world of "functions," it is possible that a function can be placed within another function. In other words, the input of one machine is another machine. We call this a composition of two functions. If we compose $f(x)$ and $g(x)$, then we get

$$f(g(x))$$

meaning that $g(x)$ is the input of f.

Example

If $f(x) = x^2 + 2$ and $g(x) = \sqrt{x}$, find $f(g(x))$.

Solution

$$f(g(x)) = f(\sqrt{x})$$
$$= (\sqrt{x})^2 + 2$$
$$= x + 2$$

---

10 Let $f(x) = x - 3$ and $g(x) = 4 - 3x$.

(a) Find $f(g(1))$. 　　　　　　　　　　(b) Find $g(f(1))$.

As one may check, the order of composition matters in this arithmetic. Now, try out the following question with a bit more abstract expression.

(c) Find $f(g(ax + b))$.

Can a function be composed by itself? Weird it may sound, but yes! For example, if $f(x) = ax + b$, then $f(f(x)) = f(ax + b) = a(ax + b) + b = a^2x + (ab + b)$.

If $f(x) = x^2 + 2$, then find $f(f(x))$.

Solution

$$\begin{aligned} f(f(x)) &= f(x^2 + 2) \\ &= (x^2 + 2)^2 + 2 \\ &= x^4 + 4x^2 + 4 + 2 \\ &= x^4 + 4x^2 + 6 \end{aligned}$$

11 If $f(x) = 2x + 10$, for what value of x does $f(f(x)) = x$?

12 If $f(x) = x - 3$ and $f(f(a)) = 4$, find the value of a.

If we wish to change the role of input and output, we may have to undo the whole process. Same goes for functions! If $f(x) = y$, then we get the machine that reverses the process of the original machinery, called the inverse of f, also known as $f^{-1}(x)$. Functions f and g are inverse to one another if and only if

- $g(f(x)) = x$ for all values of x in the domain of f.

- $f(g(x)) = x$ for all values of x in the domain of g.

Example

If $f(x) = x + 3$, find the inverse function of $f(x)$.

Solution
Use the property that $f(f^{-1}(x)) = x$ for a function $f(x)$ and its inverse function $f^{-1}(x)$.

$$f(f^{-1}(x)) = x$$
$$f^{-1}(x) + 3 = x$$
$$f^{-1}(x) = x - 3$$

13 Find the inverse of the function $f(x) = 2x - 9$.

14 The function f is the inverse of the function $g(x) = \dfrac{2x - 3}{x + 5}$. Find $f(4)$.

15 Let $f(x) = x^2$ and let the domain of f be all real numbers.

(a) Does f have an inverse?

(b) Suppose $g(x) = x^2$ and let the domain of g be all nonnegative numbers. Does the answer for (a) change?

Here is the common expression we use when there are numerical expressions laid out and we need to use inverse function properties.

$$f(\triangle) = \square \longleftrightarrow \triangle = f^{-1}(\square)$$

16 If f is a function that has an inverse and $f(2) = 5$, what is $f^{-1}(5)$?

$\boxed{1}$ Let $f(x-3) = 9x^2 + 2$ for all real values of x. What is the value of $f(2)$?

$\boxed{2}$ Suppose there is a function whose input is $\dfrac{x}{3}$ where the output is $x^2 + x + 2$, i.e., $f\left(\dfrac{x}{3}\right) = x^2 + x + 2$. If the input is $3x$, what is the output in terms of x? In short, express $f(3x)$ in terms of x. (Bonus : Try finding $f(4x)$ as well.)

3 A function $f(x)$ satisfies that $f(a) + f(b) = f(ab)$ for any positive integers a and b. If $f(2) = 5$ and $f(3) = 10$, then evaluate the value of $f(12)$.

4 If $f(x) = x^2 + x + 15$, find all possible values of a such that $f(a) = 21$.

5 Let $f(x) = 2\sqrt{2x - 7} - 5$.

(a) Find the domain of $f(x)$.

(b) Find the range of $f(x)$.

6 If $f(x) = \dfrac{x - 2}{\sqrt{x - 2}}$, then is 2 in the domain of $y = f(x)$?

7 If $f(x) = 3x$ and $g(x) = 2x - 1$, find the value of $f(g(f(1)))$.

8 Given a linear function $f(x) = x + 1$. Find the value of $f^5(2)$, where $f(f(f(f(f(x))))) = f^5(x)$.

9 Given two functions, a linear one $f(x) = 3x - 8$ and a quadratic one $g(f(x)) = 2x^2 - 5x + 5$, find the value of $g(-2)$.

10 Any linear function $y = mx + b$ has an inverse function, which can be found by $f(f^{-1}(x)) = x$. Utilize the property to find the inverse function of $f(x) = 4 - 5x$.

11 Does the function $f(x) = 2x^2 - 3$ have an inverse? If not, try to restrict the domain to find the inverse function $f^{-1}(x)$.

12 A linear function $y = mx + b$ has an inverse function if $m \neq 0$. However, when we look for the specific values of inverse functions, we do not need to find the inverse function. We can utilize $f^{-1}(a) = b$ equaling $a = f(b)$. If $f(x) = 5x - 2$, what is the value of $f^{-1}(3)$?

1 $f(2) = 227$

2 $f(3x) = 81x^2 + 9x + 2$. Likewise, $f(4x) = 144x^2 + 12x + 2$

3 $f(12) = 20$

4 $a = -3$ or $a = 2$

5

(a) $[\frac{7}{2}, \infty)$

(b) $[-5, \infty)$

6 No. 2 is not in the domain of $y = f(x)$.

7 $f(g(f(1))) = 15$

8 $f^5(2) = 7$

9 $g(-2) = 3$

10 $f^{-1}(x) = -\dfrac{x}{5} + \dfrac{4}{5}$

11 If $x \geq 0$, then $f^{-1}(x) = \sqrt{\dfrac{x+3}{2}}$. If $x \neq 0$, then $f^{-1}(x) = -\sqrt{\dfrac{x+3}{2}}$.

12 $f^{-1}(3) = 1$

Topic 13

Graphs and Transformation

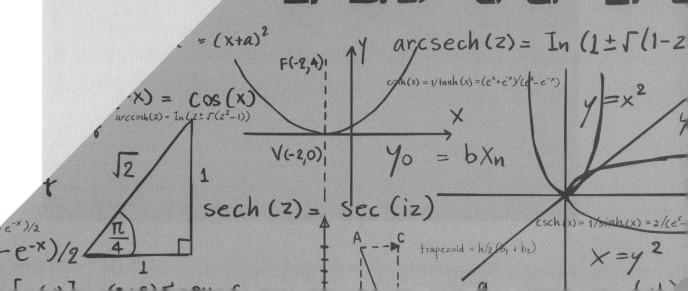

13.1 Graphing Functions

Graph of a function is a diagram of set of points (x, y) satisfying $y = f(x)$.

1 Given a quadratic function $f(x) = x^2 - 4x + 3$, solve the following questions.

(a) Graph the equation $y = f(x)$.

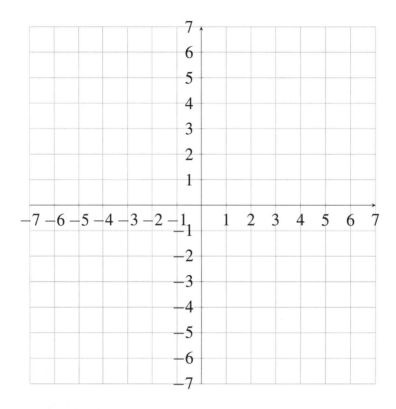

(b) Find the x-intercepts of the graph.   (c) Find the y-intercept of the graph.

2 The graph of $y = f(x)$ is given below.

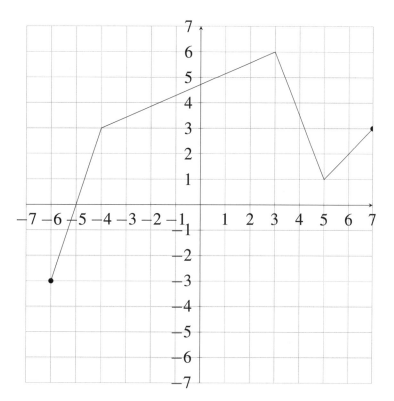

(a) Find $f(3)$ and $f(f(3))$.

(b) Find the domain of the function.

(c) Find the range of the function.

Our observations in this problem give us a general rule for determining whether or not a graph can represent a function. Specifically, we note that a graph represents a function if and only if for each value of x there is no more than one corresponding value of y. For each value of x, we have a vertical line; for example, for $x = 2$, we have the vertical line that is the graph of $x = 2$. The points on a graph with x-coordinate equal to 2 are those points where the vertical line $x = 2$ hits the graph.

A graph represents a function if and only if every vertical line passes through no more than one point on the graph. This test is called the vertical line test.

1. Draw any vertical line.

2. If it has at most one point of intersection, keep drawing vertical lines.

3. If it has at least two points of intersection, the graph fails the vertical line test.

4. If all vertical lines have at most one point of intersection with the graph of the function, then the graph passes the vertical line test.

3 Determine whether the following graphs represent functions.

(a)   (b)

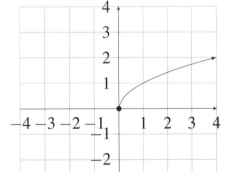
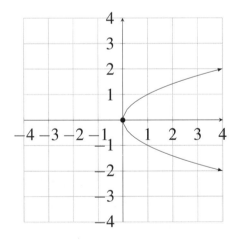

4 Sketcth the graph of $y = f(x)$.

(a) $f(x) = 1$: The graph of constant function is a horizontal line.

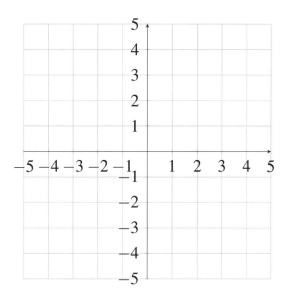

(b) $f(x) = x$: The graph of linear function is a line.

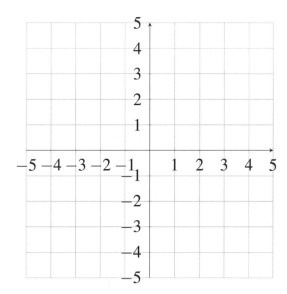

(c) $y = x^2 + 1$: The graph of quadratic function is a parabola.

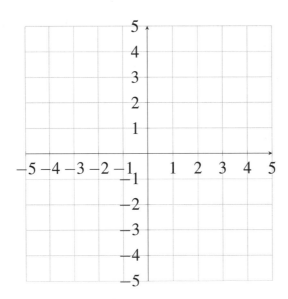

(d) $f(x) = \sqrt{x}$: The graph of square root function (or radical function) is a portion of parabola.

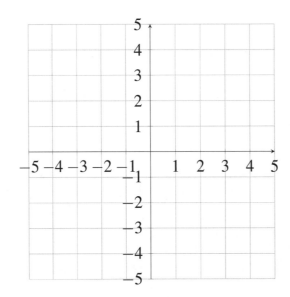

(e) $f(x) = \dfrac{1}{x}$: The graph of rational function is a hyperbola.

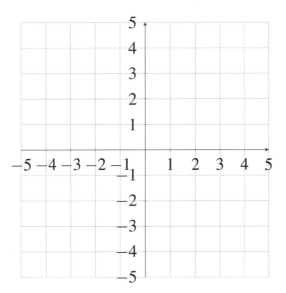

5 Determine whether the following graph is a function, and state the domain and range.

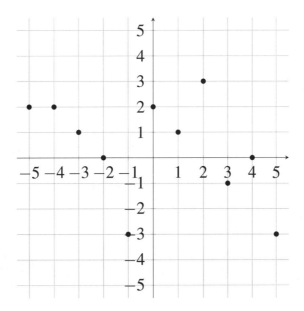

13.2 Transformation

There are four types of transformation : dilation, reflection, translation, and rotation. In this section, we only cover the first three types. Let's learn how to dilate the graphs! Multiplying a positive constant to either input or output has the effect of dilation.

- $y = f(kx)$ horizontally shrinks(or strecthes) the graph of $y = f(x)$ by k unit.

- $y = kf(x)$ vertically strecthes (or shrinks) the graph of $y = f(x)$ by k unit.

$\boxed{6}$ The following graph of $y = f(x)$ will be dilated either vertically or horizontally. Graph the resulting graph in the following xy planes.

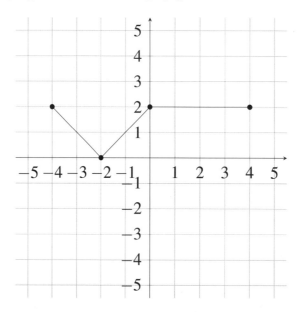

(a) $y = f(2x)$

(b) $y = 2f(x)$

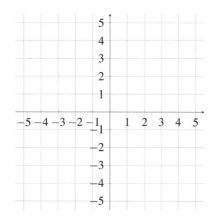

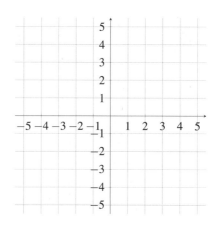

If we multiply a fraction to input or output, then it is the opposite dilation of the previous one.[1] Let's have a look at the following example.

7 The following graph of $y = f(x)$ will be dilated either vertically or horizontally. Graph the resulting graph in the following xy planes.

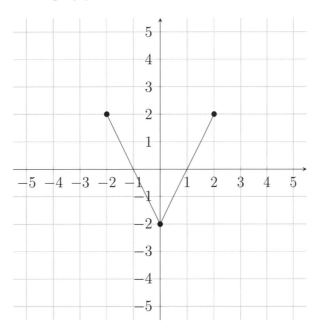

(a) $y = f(\frac{1}{2}x)$ (b) $y = \frac{1}{2}f(x)$

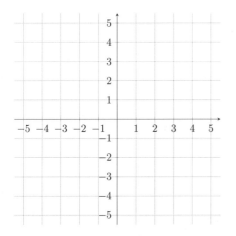   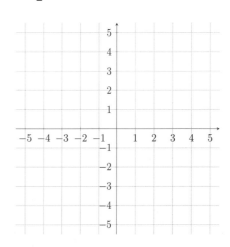

---

[1]If $y = f(kx)$, then $1 < k$ means horizontal **shrinking**, whereas $0 < k < 1$ means horizontal **stretching**. On the other hand, if $y = kf(x)$, then $1 < k$ means vertical **stretching**, while $0 < k < 1$ means vertical **shrinking**.

Multiplying negative constant to input or output has the effect of reflecting the original graph in which there are three major types of reflection : x-axis, y-axis and the origin.

- $y = f(-x)$: Reflection about the y-axis.

- $y = -f(x)$: Reflection about the x-axis.

- $y = -f(-x)$: Reflection about the origin.

8 The following graph of $y = f(x)$ will be reflected about the directions given below. Graph the resulting graph in the following xy planes.

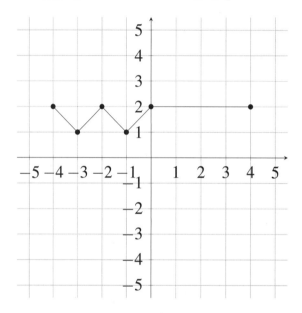

(a) $y = f(-x)$

(b) $y = -f(x)$

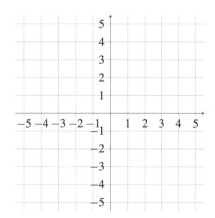

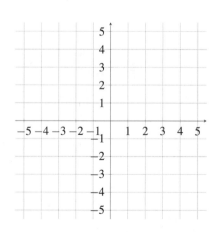

If we add a constant to input or output, the effect of the transformation is known as translation (or shift).

- $y = f(x) + k$: Shift it vertically by k units.

- $y = f(x + k)$: Shift it horizontally by k units.

9 The following graph of $y = f(x)$ will be translated(or shifted) about the directions given below. Graph the resulting graph in the following xy planes.

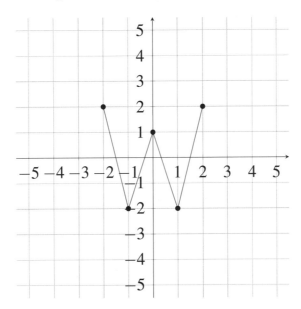

(a) $y = f(x) + 3$

(b) $y = f(x - 2)$

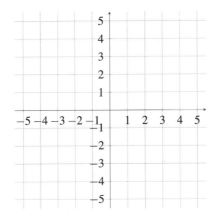

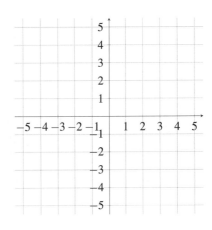

10 The graph of $y = f(x)$ is shown below. Sketch the graph of $y = 2f(x-2) - 1$ on the same diagram.

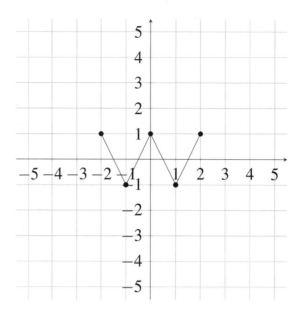

11 Suppose f is a function such that $f(2) = 4$. Find a point that must be on the graph of

(a) $y = f(2x) + 2$ (b) $y = f(-x) + 1$

(c) $y = f\left(\dfrac{x}{2}\right) - 1$ (d) $y = 2f(2x-4) + 4$

13.3 Graphs of Inverse Functions

The graph of the original function $y = f(x)$ and that of its inverse $y = f^{-1}(x)$ have an interesting symmetry. We will have a look at what this is.

12 Let $f(x) = 2x + 1$.

(a) Find $f^{-1}(x)$.[2]

(b) Graph $y = f(x)$, $y = f^{-1}(x)$ and $y = x$.

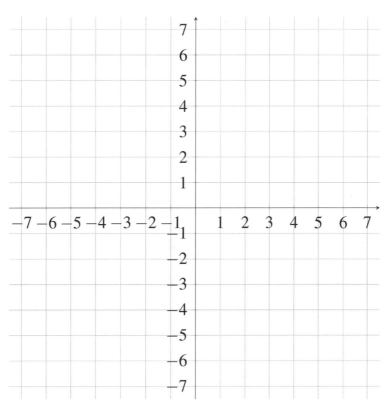

---

[2]Recall from the previous chapter that $f(f^{-1}(x)) = x$.

Now, we notice that the graph of $y = f(x)$ and of $y = f^{-1}(x)$ are symmetric about the line $y = x$.

- If (a,b) is on the graph of $y = f(x)$, then (b,a) is on the graph of $y = f^{-1}(x)$.

- The intersection point(s) between the graph of $y = f(x)$ and $y = f^{-1}(x)$ are not reflected about the line $y = x$. In other words, it is **invariant** point.

13 If $(1,3)$, $(2,5)$, $(3,0)$, and $(4,4)$ are on the graph of $y = f(x)$, state four points on the graph of $y = f^{-1}(x)$. Plot the points and see if the midpoints between the original point and the corresponding point on the inverse function are on the graph of $y = x$.

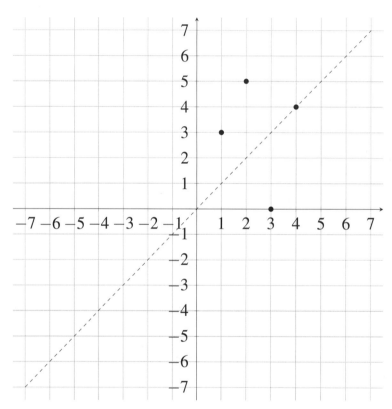

In the previous section, VLT determines whether a graph represents a function. Is there a similar test for the graph of a function to see whether it has an inverse function or not?

Specifically, we note that a function has an inverse if and only if it is reversible. That is, a function has an inverse if we can tell from the output of the function what the input was. If f has an inverse, then when we graph $y = f(x)$, there is only one point on the graph with the same y-coordinate. Otherwise, two different inputs (the x-coordinates of points on the graph) produce the same output (the common y-coordinate) for the function, so it couldn't have an inverse. Therefore, we draw any horizontal line and see if the line intersects with the graph more than one point. This is called a horizontal line test, also known as HLT, to find out whether or not a function has an inverse.

Example

Explain why $y = x^2$ has no inverse function.

Solution
Inverse function exists if and only if the function is 1-to-1. On the other hand, $y = x^2$ is 2-to-1, meaning that 2 values of x are connected to 1 value of y. Hence, if you reverse the role of x and y, then $x = y^2$ will have 1 value of x attached to 2 distinct values of y. This implies that $x = y^2$ is not a function.

14 Apply a horizontal line test to the graphs of the following functions to see if they have inverse functions.

(a)

(b)

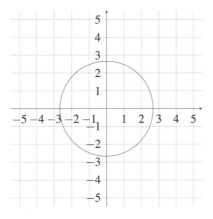

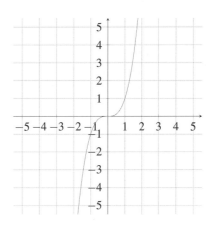

15 The graph of $y = f(x)$ is shown below. Draw the graph of $y = f^{-1}(x)$. In fact, this is the graph of $y = x^{1/3} = \sqrt[3]{x}$.

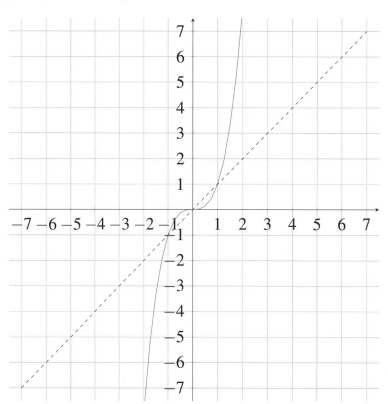

1 Given a quadratic function $f(x) = (x-1)^2 - 1$, solve the following questions.

(a) Graph the equation $y = f(x)$.

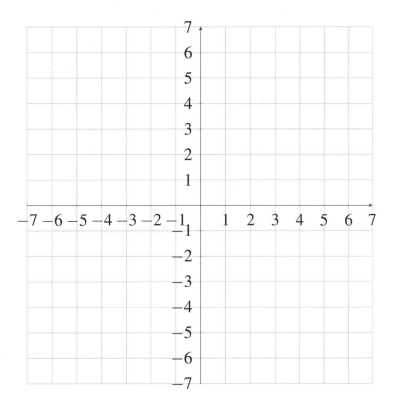

(b) Find two x-intercepts of the graph.

(c) Find the y-intercept of the grpah.

2 The graph of $y = f(x)$ is given below.

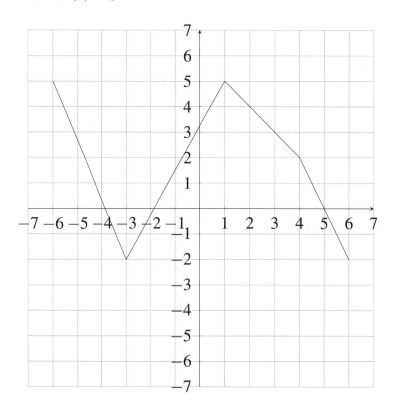

(a) Find $f(2)$.

(b) Find $f(f(4))$.

(c) Find the number of x such that $f(x) = 3$.

(d) State whether the function $y = f(x)$ has an inverse function.

3 The following graph of $y = f(x)$ will be translated(or shifted) about the directions given below. Graph the resulting graph in the following xy planes.

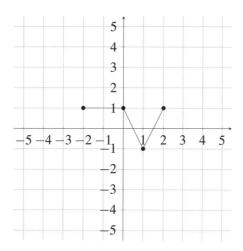

(a) $y = 2f(-x) + 1$

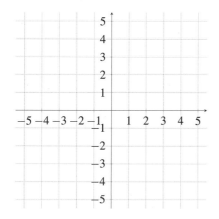

(b) $y = -f(2-x) + 1$

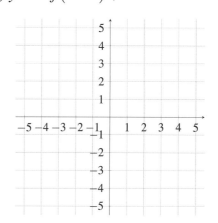

(c) $y = f(2x-2) - 1$

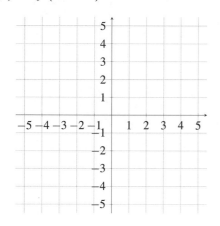

(d) $y = -f(-x+1) + 1$

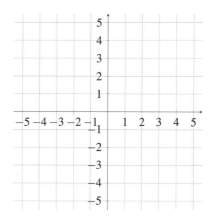

4 Assume $(1,3)$ is on the graph of $y = f(x)$. Find a point that must be on the following graphs.

(a) $y = f(-x+2)+2$

(b) $y = f(x-1)+2$

(c) $y = -f(3-x)-3$

(d) $y = 2f(2x)-2$

5 The function $f(x) = mx+b$ is such that f and f^{-1} are the same function. Find all possible values of m.

6 The function f satisfies $f(x) = 2$ for $-2 \le x < 1$. If $y = f(x)$ satisfies

$$\frac{f(x-3)}{2} = f(x)$$

for all real x, draw the portion of the graph of $y = f(x)$ for which $-5 \le x < 7$.

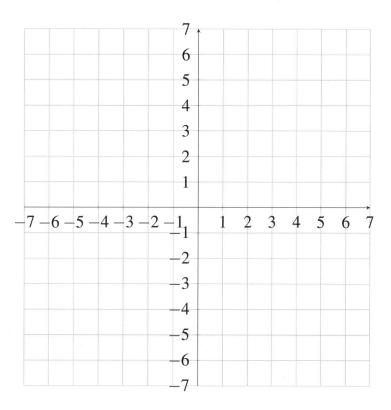

1

(a)

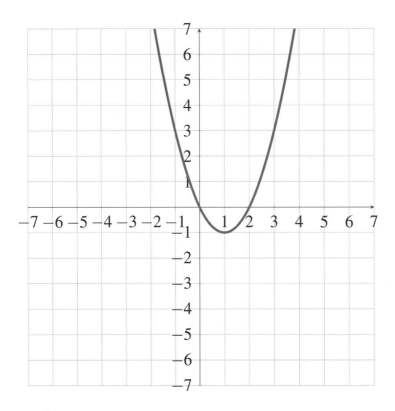

(b) The x-intercepts are 0 and 2.

(c) The y-intercept is 0.

2

(a) $f(2) = 4$

(b) $f(f(4)) = 4$

(c) There are 3 x's satisfying $f(x) = 3$.

(d) It does not pass HLT, so the function $y = f(x)$ has no inverse function.

(a) $y = 2f(-x) + 1$

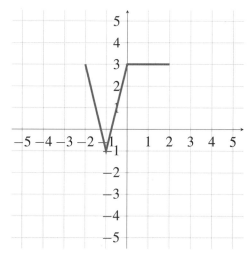

(b) $y = -f(2-x) + 1$

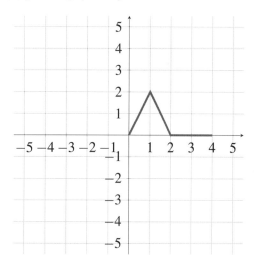

(c) $y = f(2x-2) - 1$

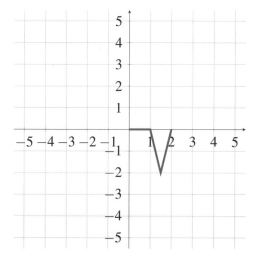

(d) $y = -f(-x+1) + 1$

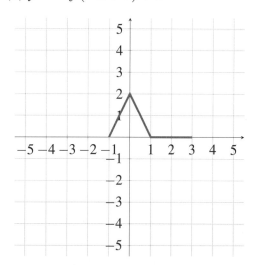

(a) $(1, 5)$

(b) $(2, 5)$

(c) $(2, -6)$

(d) $\left(\dfrac{1}{2}, 4 \right)$

5 $m = \pm 1$

6

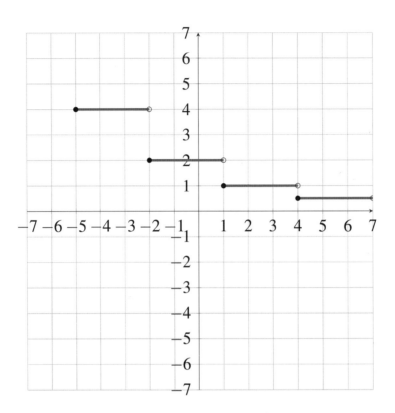

Topic 14

Polynomial Arithmetic

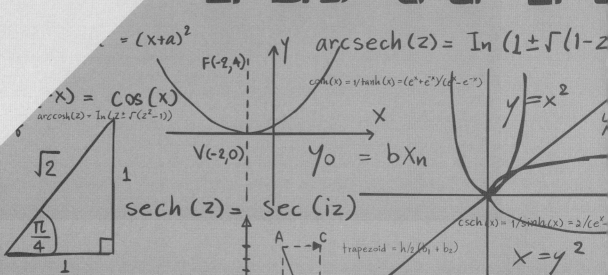

14.1 Polynomials

A polynomial of one variable is a sum of terms in which each term is a constant times a variable raised to a nonnegative integer power. The following are all polynomials:

$$x^5 + 3x^2 + 1, \qquad t^9 - 3t^8 + 276t, \qquad -8z^{10} + z^5 - 1.$$

As the example polynomials above illustrate, we usually write polynomials such that the exponents of the variable decrease from left to right. Polynomials can also have more than one variable. For example, these are also polynomials:

$$2xy, \qquad \frac{x^3}{8} - \frac{y^3}{8}, \qquad 3z^2y - 5zy + 3zy^2 + 2.$$

All variables in polynomials must have nonnegative powers, and the variables can't be in denominators or under square root signs, etc. The following are not polynomials:

$$x + \frac{1}{x}, \qquad x^2 + \frac{x}{y} - 3y^2, \qquad \sqrt{a^2 + b^2}.$$

Nearly all the work we do in this book will be with polynomials that have only one variable. We call the highest power of the variable in such a polynomial the degree of the polynomial, and we call the term containing this highest power the leading term of the polynomial. For example, the leading term of

$$f(x) = 3x^4 + 2x^2 - 7$$

is $3x^4$. To denote that f has degree 4, we write $\deg f = 4$.

1 Let $f(x) = x^3 - 4x + 7$ and $g(x) = -3x^3 + 2x^2 + x - 7$.

(a) Find $f(x) + g(x)$. (b) Find $f(x) - g(x)$.

Multiplying two polynomials results in a polynomial. What matters is the distribution property.

$\boxed{2}$ If $p(x) = 3x^2 - 2x + 3$ and $q(x) = x^3 - 2x^2 + x - 7$, find the product

$$p(x) \cdot q(x) = (3x^2 - 2x + 3)(x^3 - 2x^2 + x - 7)$$

$\boxed{3}$ Find either one of the two : the sum $a^2 + b^2$ or $(a+b)^2$, where a and b are constants from $(x^3 + bx^2 - 7x + 9)(x^2 + ax + 5)$ where the product contains $x^5 + 13x^4 + 38x^3 + \cdots + 45$.

Find the coefficient of x^3 for $(x^2 + 2x + 3)(x^2 - x - 7)$.

Solution
Instead of expanding the product, we can specifically look for the terms with x^3. We get $x^2(-x) + 2x(x^2) = -x^3 + 2x^3 = x^3$. Hence, the coefficient of x^3 is 1.

$\boxed{4}$ Given a product of two polynomial expressions, $(x^2 - 3x + 4)(2x^2 + ax + 7)$, the only term that we know is the coefficient of x^3, which is -11. Find the value of a.

14.2 Polynomial Division

A rational expression of polynomials can also be added, subtracted, multiplied or divided. First, let's learn how to simplify a given rational expression.

5 Simplify the following rational expressions.

(a) $\dfrac{x-2}{3x^2-12}$

(d) $\dfrac{x^3-x^2-42x}{2x^2-20x+42}$

(b) $\dfrac{x-1}{x^2-5x+4}$

(e) $\dfrac{9x^2-81x}{x^3-8x^2-9x}$

(c) $\dfrac{x^2-5x-14}{x^2+4x+4}$

(f) $\dfrac{x^2-2x-8}{2x^3-24x^2+64x}$

In order to find out the excluded values for the rational expressions, we look at the denominator. However, there are two types of excluded values that could appear in the process of simplification.

- Canceled terms : the zeros of the canceled terms turn into holes in the graph of the rational expression(or function).

- Remaining terms : the zeros of the remaining terms in the denominator turn into vertical asymptotes[1].

6 State the excluded values for each of the following expressions.

(a) $\dfrac{14}{3x-1}$

(d) $\dfrac{x^2+3x+2}{x^2+1}$

(b) $\dfrac{24x-12}{2x-1}$

(e) $\dfrac{5x+7}{25x^2-49}$

(c) $\dfrac{3x^2}{x^2-9}$

(f) $\dfrac{52x-12}{4}$

---

[1]Vertical asymptotes are the vertical lines in the graph that do not intersect with the graph, whereas the graph approaches that line as x gets closer to the vertical lines.

$$P(x) = (x-a) \times Q(x) + P(a)$$

Dividend Divisor Quotient Remainder

7 Divide.

(a) $x^3 - 2x^2 - 14x - 5$ by $x + 3$

(c) $x^3 + 5x^2 - 16x + 7$ by $x - 4$

(b) $x^3 - 4x^2 - 30x + 18$ by $x - 2$

(d) $12x^3 - 4x^2 + 5x + 2$ by $2x - 1$

When we add or subtract rational expressions, we need to find the least common denominator to equalize the denominators.

$$\frac{a}{b} + \frac{c}{d} = \frac{ad}{bd} + \frac{bc}{bd} = \frac{ad + bc}{bd}$$

8 Simplify each of the following expressions.

(a) $\dfrac{2}{3x+3} + \dfrac{1}{x-4}$

(c) $\dfrac{4x}{x-3} + \dfrac{3}{x+5}$

(b) $\dfrac{5}{x-1} + \dfrac{4}{2-x}$

(d) $\dfrac{5}{2-x} - \dfrac{12x}{x^2-4}$

14.3 Rational Equations

Combining two rational expressions into a single fraction, we can solve rational equations that consist of polynomials, especially when the degree of the resulting polynomials is at most 2.

9 Solve for x.

(a) $\dfrac{6}{x^2 - 4x + 4} = \dfrac{1}{x^2 - 4x + 4} - \dfrac{1}{x - 2}$

(c) $\dfrac{x + 3}{x^2 - 3x} = \dfrac{1}{x - 3} - \dfrac{x + 5}{x^2 - 3x}$

(b) $\dfrac{3}{x + 3} + \dfrac{3x}{x + 3} = 1$

(d) $\dfrac{4}{x} = \dfrac{1}{x - 4} - \dfrac{x + 2}{x^2 - 4x}$

10 Solve the following questions.

(a) Solve for x if $\dfrac{x+2}{x} + \dfrac{4x+2}{x^2-3x} = 1 - \dfrac{1}{x}$.

(b) Solve for x if $\dfrac{1}{x} - 5 = \dfrac{1}{x^2+x}$.

(c) Find a polynomial (with the leading coefficient of 2) that has the same zero for
$\dfrac{1}{2} + \dfrac{1}{2x} = \dfrac{x^2-7x+10}{x-1}$.

1 Let $p(x) = 2x - 5$ and $q(x) = 3x^2 + 7x - 4$.

(a) Find $2p(x) + 3q(x)$.

(b) Find $p(x) \cdot q(x)$.

2 Given a quartic polynomial with integer coefficients,

$$f(x) = x^4 - 3x^3 + ax^2 + x - 2$$

where a is a constant, if $f(3) = 2$, then what is the value of a?

3 Simplify the following rational expressions into a single term.

(a) $\dfrac{1}{x-1} - \dfrac{2x-3}{x-3}$

(b) $\dfrac{1}{x^2-4x} - \dfrac{x-4}{x}$

4 Divide.

(a) $x^4 - x^3 + x^2 - x + 1$ by $x+1$

(b) $x^5 - 1$ by $x-1$

5 Solve for x.

(a) $\dfrac{x^2 - x - 6}{x^2} = \dfrac{x - 6}{2x} + \dfrac{2x + 12}{x}$

(b) $\dfrac{x^2 - 2x + 8}{3x^2} = \dfrac{1}{3} + \dfrac{1}{3x}$

(c) $\dfrac{2}{x} - \dfrac{2}{x^2 + x} = 10$

$\boxed{1}$

(a) $9x^2 + 25x - 22$

(b) $6x^3 - x^2 - 43x + 20$

$\boxed{2}$ $a = \dfrac{1}{9}$

$\boxed{3}$

(a) $\dfrac{-2x^2 + 6x - 6}{x^2 - 4x + 3}$

(b) $\dfrac{-x^2 + 8x - 15}{x^2 - 4x}$

$\boxed{4}$

(a) $x^4 - x^3 + x^2 - x + 1 = (x + 1)(x^3 - 2x^2 + 3x - 4) + 5$

(b) $x^5 - 1 = (x - 1)(x^4 + x^3 + x^2 + x + 1)$

$\boxed{5}$

(a) $x = -\dfrac{2}{3}, -6$

(b) $x = \dfrac{8}{3}$

(c) $x = -\dfrac{4}{5}$

Topic 15

Statistics

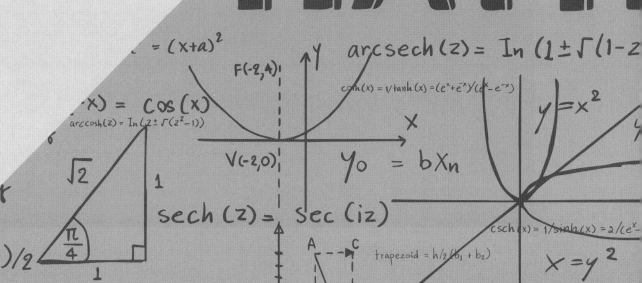

15.1 Measures of Central Tendency

Given a list of data sets, it is best for us to find the central measures to see how data is distributed around the center. There are three measures in which we can utilize for analysis.

- The mode is the most frequent value.

- The median is the middle value when data is listed from least to greatest.

- The mean is the arithmetic average.

Example

The arithmetic mean of 12 scores is 82. When the highest and lowest scores are removed, the new mean becomes 84. If the highest of the 12 scores is 98, what is the lowest score?

Solution
Since the arithmetic mean is 82, the total sum of values is 12×82. Since two values are removed, the new total sum is now 10×84. Since the highest value is 98, if we call the lowest score as x, then $x + 98 = 984 - 840 = 144$. Hence, $x = 144 - 98 = 46$.

1 What is the mean, median and mode of $\{1, 2, 3, 3, 4, 5, 6\}$?

15.2 Measures of Spread

In Prealgebra and the previous section, we went over the basic concepts of the measures of central tendency - mean, median, and mode - in detail. In this section, we will learn about the measures of spread, especially standard deviation. The standard deviation is a measure that captures how much each data deviates from the mean value. In other words, we need *mean* to compute the exact value of standard deviation.

Given n list of data sets, $\{x_1, x_2, x_3, \cdots, x_n\}$, if the mean value is $\overline{X}$, then the standard deviation, denoted by σ, is given by

$$\sigma = \sqrt{\frac{(x_1 - \overline{X})^2 + (x_2 - \overline{X})^2 + \cdots + (x_n - \overline{X})^2}{n}}$$

On the other hand, another measure of spread is given by the range and the interquartile range, also known as IQR. Given a median, the lower half of the data also has its median, known as the Lower Quartile. Likewise, the upper half of the data has its median, a.k.a the Upper Quartile. The interquartile range is the difference between UQ and LQ, whereas the range is the difference between the maximum and minimum.

Example

Given a data list of $\{2018, 2020, 2022, 2024\}$, find the following measures of spread.
(a) standard deviation   (b) range   (c) interquartile range

Solution
(a) The mean value is given by $\dfrac{2018 + 2020 + 2022 + 2024}{4} = 2021$.
Hence, the standard deviation can be computed by

$$\sqrt{\frac{(2018 - 2021)^2 + (2020 - 2021)^2 + (2022 - 2021)^2 + (2024 - 2021)^2}{4}}$$

which can be simplified by $\sqrt{5}$.
(b) The range is $2024 - 2018 = 6$.
(c) The interquartile range is $2023 - 2019 = 4$.

2 Compute the standard deviation of $\{1,2,3,4,5\}$. The square of standard deviation is known as *variance*. It is best to find the variance, then take a square root of the value.

3 Find the range and interquartile range of $\{2,3,3,4,5,6\}$.

4 What is the IQR of the following data set?

$$\{2019, 2020, 2021, 2022, 2022, 2023\}$$

15.3 Line of Best Fit

The line of best fit, also known as a linear regression, shows an association between two variables. An association is a relationship between two numerical variables that can be displayed on a scatterplot. In statistics, the line of best fit is used to predict or model what happens in the data set.

A graph that displays two numerical variables, which can be described by its direction, strength, form and outliers is called scatterplot.

- Direction : positive or negative, describing the pattern in a scatterplot and can be confirmed by looking at the slope of the Least-Square Regression Line.

- Strength : strong, moderate, or weak, describing the pattern in a scatterplot and can be checked by the amount of spread away from the Least-Square Regression Line.

- Form : linear or nonlinear, describing the pattern in a scatterplot and can be confirmed by looking at the residual plot.

- Outlier : A point on the graph that does not fit into the overall pattern or is extremely far from the rest of the data.

Here, the Least-Square Regression Line is a unique line through a scatterplot that is created by minimizing the squares of the residuals.

- Slope : it describes rate or growth, using $\dfrac{\triangle y}{\triangle x}$.

- y-intercept : it describes the initial amount.

Using the LSRL, we get the predicted value. There might be a difference between the predicted value and the actual value, which can be estimated by the distance between the actual y-value and the predicted y-value, represented by a vertical line between a dot and the LSRL.

Residual plot is a graph that helps determine whether a linear model is good or if a curved model is good. If the plot has no distinguishing pattern, then we say we have a random scatter.

5 Compare the strength from greatest to weakest.

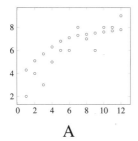

A

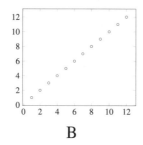

B

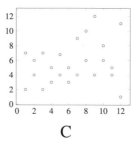

C

1 Compare the mean and median of the following list of data.

(a) $\{1, 2, 3, 4, 4, 4, 5, 6, 7\}$

(b) $\{2, 3, 5, 5, 2000\}$

2 Compare the following list of data in terms of standard deviation from least to greatest.

- $A = \{1, 3, 3, 3, 5\}$
- $B = \{2, 3, 3, 3, 4\}$
- $C = \{1, 2, 3, 4, 5\}$

3 Which of the following has the strongest positive correlation?

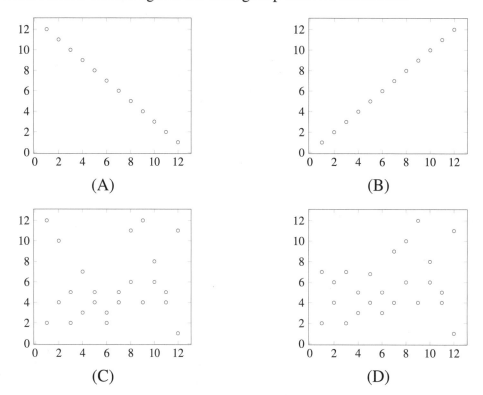

(A) (B) (C) (D)

4 Given $\{1, 2, 3, 4, 5, 20\}$, how much would mean change and median change if 20 is deleted?

5 If the line LSRL is given by $y = 100x + 3$, is the data correlated in (positive / negative) direction?

6 Given the following data plots, explain why a non-linear model fits better in terms of residuals.

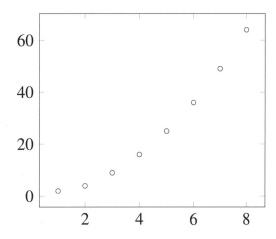

☐1

(a) The mean value is 4, and the median is 4.

(b) The mean value is 403, and the median is 5.

☐2 B, A, and C.

☐3 (B)

☐4 The mean value changes from $\dfrac{35}{6}$ to 3, whereas the median changes from 3.5 to 3.

☐5 The data is correlated in positive direction because the slope is positive.

☐6 Non-linear regression curve fits better because it has smaller residuals compared to the linear regression line.

Solution Manual

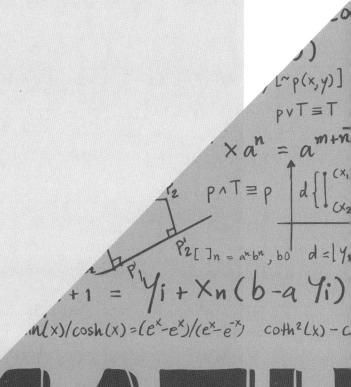

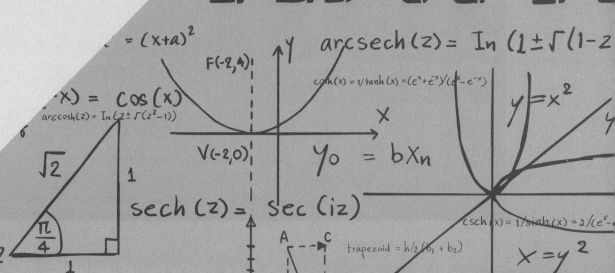

Solution for Topic 1 Questions

1

(a) $\dfrac{x+12}{x} = \dfrac{x}{x} + \dfrac{12}{x} = 1 + \dfrac{12}{6} = 1 + 2 = 3.$

(b) $3x^2 = 3(6)^2 = 3(36) = 108.$

(c) $\sqrt{5x-5} = \sqrt{30-5} = \sqrt{25} = 5.$

2

(a) $(4x-3)+(6x-7) = 10x-10(=10(x-1)).$

(b) $(2-5x)+(-17x-27) = -22x-25 = -(22x+25)$

3

(a) $a^3 \times a^4 = a^7$

(b) $(8x^3) \times (18x^2) = 144x^5$

(c) $(2x^3)^4 = 2^4 \times (x^3)^4 = 16x^{12}$

4 $-25a^3 + 40 = -5(5a^3 - 8) = 5(-5a^3 + 8) = 5(8 - 5a^3).$

5 $x(3x+2)+2(3x+2) = (x+2)(3x+2).$

6 $2\left(\dfrac{2x+48}{4} - 7\right) - x = 10$

7

(a) $6a^2 - 24a = 6a(a-4)$

(b) $4x^2 - 18x = 2x(2x-9)$

(c) $-8t^2 - 4t = -4t(2t+1) = 4t(-2t-1)$

(d) $6x^3 - 12x^2 + 12 = 6(x^3 - 2x^2 + 2)$

8 $4x(x-2) + 7(x-2) = (4x+7)(x-2)$

9

(a) $\dfrac{2}{x} + \dfrac{3x}{4} = \dfrac{8+3x^2}{4x}$

(b) $\dfrac{4}{3x} - \dfrac{5-x}{6x^2} = \dfrac{9x-5}{6x^2}$

10 $\dfrac{2}{x} + \dfrac{7x}{x+3} = \dfrac{7x^2+2x+6}{x^2+3x}$

1

(a)

$$5\sqrt{x} - 2 = 28 - \sqrt{x}$$
$$6\sqrt{x} = 30$$
$$\sqrt{x} = 5$$
$$x = 25$$

(b)

$$\frac{6}{x} + 3 = 7 - \frac{2}{x}$$
$$\frac{8}{x} = 4$$
$$\frac{8}{4} = 2$$

2

(a)

$$\sqrt[3]{1 - 2t} + 3 + 2\sqrt[3]{1 - 2t} = 6$$
$$3\sqrt[3]{1 - 2t} = 3$$
$$\sqrt[3]{1 - 2t} = 1$$
$$1 - 2t = 1$$
$$2t = 0$$
$$t = 0$$

(b)

$$\frac{x}{x+1} + \frac{3}{4} = \frac{3}{x+1}$$
$$\frac{x-3}{x+1} = \frac{-3}{4}$$
$$4(x-3) = -3(x+1)$$
$$4x - 12 = -3x - 3$$
$$7x = 9$$
$$x = \frac{9}{7}$$

3

$$(x-3)^2 + 4 = 6 - (x-3)^2$$
$$2(x-3)^2 = 2$$
$$(x-3)^2 = 1$$
$$x - 3 = \pm 1$$
$$x = 4 \text{ or } 2$$

4

(a) $r^2 + 2rs + s^2 = 1$

(b) $(r+s)^2 = 1$

(c) $r^2 + 3r^2s + 3rs^2 + s^3 = 1$

(d) $(r+s)^3 = 1$

$\boxed{5}$

(a) $(x + y - z) + (2x - 3y + 11z) = 3x - 2y + 10z$

(b)

$$4ab + 3cd + 2cd - 11ab + 4 - 1 = (4ab - 11ab) + (3cd + 2cd) + (4 - 1)$$
$$= -7ab + 5cd + 3$$

$\boxed{6}$

$$(3xy^2) \cdot (2xy^6) = (3 \times 2) \cdot (x \times x) \cdot (y^2 \times y^6)$$
$$= 6 \cdot x^2 \cdot y^8$$
$$= 6x^2 y^8$$

$\boxed{7}$

(a) $\dfrac{28x^3 y^2}{21x^4 y^6} = \dfrac{4}{3xy^4}$

(b) $\dfrac{-4x^2 y^3 z^7}{-16x^3 y^7 z^2} = \dfrac{z^5}{4xy^4}$

$\boxed{8}$ $\dfrac{18x^5 y^6}{6x^2 y^3} = 3x^3 y^3$

$\boxed{9}$ $(a + b)(2a + 3b) = 2a^2 + 5ab + 3b^2$

$\boxed{10}$

(a) $(2a + 3b) - (4a + 6b) = (2a - 4a) + (3b - 6b) = -2a - 3b = -(2a + 3b)$

(b) $4(x - y + z) - 3(2x - 2y + 2z) = 4x - 4y + 4z - 6x + 6y - 6z = -2x + 2y - 2z = -2(x - y + z) = 2(-x + y - z)$

$\boxed{11}$

(a) $6x^2 + 8xz = 2x(3x + 4z)$

(b) $7a^2 b^2 - 21ab^3 + 14a^2 b^3 = 7ab^2(a - 3b + 2ab)$

$\boxed{12}$ $\dfrac{2x + 4y}{8} \times \dfrac{3xy}{x^2 + 2xy} = \dfrac{x + 2y}{2} \times \dfrac{3y}{x + 2y} = \dfrac{3y}{2}$

13

(a)

$$\frac{5y}{6x^2} - \frac{4}{3xy} = \frac{5y^2}{6x^2y} - \frac{8x}{6x^2y}$$
$$= \frac{5y^2 - 8x}{6x^2y}$$

(b)

$$\frac{2a^3}{a^3b} + \frac{3b}{a} - \frac{3b-3}{6ab-6a} = \frac{2}{b} + \frac{3b}{a} - \frac{3(b-1)}{6a(b-1)}$$
$$= \frac{2}{b} + \frac{3b}{a} - \frac{1}{2a}$$
$$= \frac{4a + 6b^2 - b}{2ab}$$

14

$$xy - 2x - 3y = 2$$
$$xy - 2x - 3y + 6 = 8$$
$$(x-3)(y-2) = 8$$

Hence, $(x,y) = (11,3),(7,4),(5,6),(4,10)$.

1. $(x, y) = (1, \frac{1}{3})$, $(0, -\frac{1}{3})$, $(2, 1)$.

2. $(x, y) = (2, 1)$

3. $(p, q) = (2, -8)$

4. $(x, y) = (673, 173)$

5. $(x, y) = (3, -2)$

6. $(a, b, c) = (1, 2, 3)$

7. 15 quarters

8. $a + b = 64$

9. The sum of their weights is $\frac{400}{7}$. Hence, Eric's weight is 25.

10. $12g + 12r = 34$

$\boxed{1}$ Follow the following steps.

#1. Let's set up the direct proportion ratio between the height and the shadow length, i.e., 5 : 10.

#2. Then, we set up another equation $5 : 10 = x : 32$, where x is the length of the flagpole.

#3. Hence, $x = 16$ feet.

$\boxed{2}$ Follow the following steps.

#1. Work rate is inversely proportional.

#2. Total amount of work will not change, so 12×18 does not change.

#3. We set up an equation $12 \times 18 = 9 \times x$, where x is the number of hours that nine people have to work.

#4. Hence, $x = 24$ hours.

$\boxed{3}$ Follow the following steps.

#1. Paraphrase the first sentence into $x = \dfrac{k}{y^2}$.

#2. Substitute $x = 9$ and $y = 2$ to retrieve $k = 36$.

#3. Hence, $x = \dfrac{9}{25}$ if $y = 10$.

$\boxed{4}$ Follow the following steps.

#1. Paraphrase the given condition such that Let $a = kbc$ where $k = \dfrac{1}{6}$.

#2. Substitute the given expressions. Hence, if $b = 4$ and $c = 9$, then, $a = 6$.

$\boxed{5}$ Follow the following steps.

#1. Let t be the amount of time Bob reads.

#2. $90t = 60(t + \dfrac{1}{2})$, so $t = 1$ (hour).

#3. Bob catches upto Amy's page at 2 : 30.

$\boxed{6}$ First, $20t = 60$ where t is the amount of time Amy and Bob traveled, and the dog must be running for 30 km per hour. Hence, the poor dog must be running for 90 km.

$\boxed{7}$ Using the rate equations, we get 2 hours and 55 minutes, i.e, 175 minutes.

$\boxed{8}$ Let x be the speed of water current, and y the speed of the boat in still water. Hence, $x + y = 200$ and $y - x = 125$. Therefore, $x = \dfrac{75}{2}$.

1

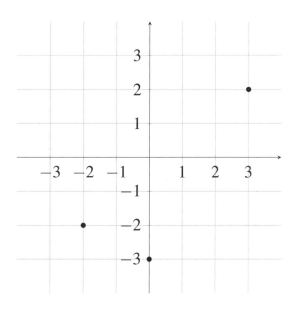

2 The distance between the two points is

$$\sqrt{(-2-3)^2+(-3-2)^2}=\sqrt{25+25}=\sqrt{50}=5\sqrt{2}$$

3 Instead of using the distance formula, we simply take the difference of the x-coordinates. Hence, the distance between the two points is 8.

4 (C) has the distance of 5, so (C) is the farthest from the origin.

5 Given $x-y=1$, $(x,y)=(0,-1)$, $(1,0)$, and $(2,1)$.

6 Since $x+2y=12$, the rate of change of y per change of x equals $-\dfrac{1}{2}$, which is invariant.

7 If the line has a negative slope, then it means we have either $\dfrac{+}{-}$ or $\dfrac{-}{+}$. Interpreting the meaning of these two fractions with signs, we get that the graph must be falling as x increases.

8 The slope of the line passing through $(-3,5)$ and $(2,-5)$ equals

$$\frac{\triangle y}{\triangle x}=\frac{5-(-5)}{-3-2}=\frac{10}{-5}=2$$

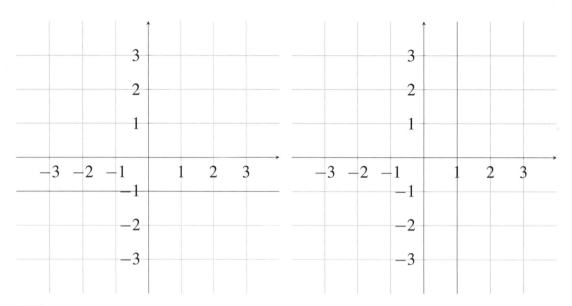

9

(a)   (b)

10 Follow the following steps.

#1. Choose any point on the graph, and move to the right direction.

#2. In order to reach the graph back agan, we need to go downward.

#3. This means that $\dfrac{\triangle y}{\triangle x} = \dfrac{-}{+}$.

#4.The slope must be negative because the graph is falling.

11

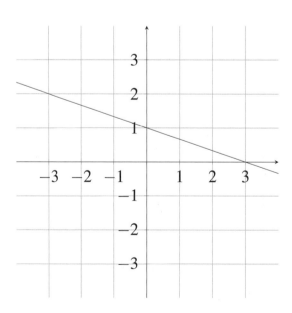

$\boxed{12}$ A, B, and C are not collinear because the slope of $\overline{AB}$, which is $\dfrac{-13}{8}$, and that of $\overline{BC}$, which is $\dfrac{-3}{2}$, are different.

$\boxed{13}$

(a) The midpoint of PQ is $\left(-\dfrac{5}{2}, 1\right)$.

(b) The coordinates of T must be $(-4, 0)$.

$\boxed{14}$

(a) $(3, 2)$, $(0, 0)$, and $(-3, -2)$.

(b) The slope between (x, y) and $(3, 2)$ is $\dfrac{y-2}{x-3}$.

(c) $2x + (-3)y = 0$ where $A = 2$, $B = -3$ and $C = 0$.

$\boxed{15}$

(a) The slope of the line is $\dfrac{-7}{5}$.

(b) $y = -\dfrac{7}{5}x + 4$

$\boxed{16}$

(a) $y = -2x + 7$

(b) $y = 2$

$\boxed{17}$ $y = \dfrac{3}{4}x - \dfrac{7}{2}$, so $3x - 4y = 14$.

$\boxed{18}$ The line equation must pass through $(-1, -5)$ and $(4, 5)$. Hence, $y = 2x - 3$. The standard form of the line must be $2x - y = 3$.

$\boxed{19}$

(a) Two points on the line are $(0, 1)$ and $(2, 0)$.

(b) The slope of the line is $\dfrac{-1}{2}$

(c)

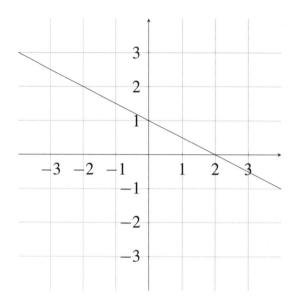

20 The triangle formed by the lines and the x-axis has the base of 6 with the height of 20, so the area must be 60.

21 Let $A(3t, 1)$, $B(4t, 2)$ and $C(100 - 6t, 3)$ Then, in order for A, B, and C to be collinear, the slope values must be invariant. Hence, the slope of $\overline{AB}$ is given by $\dfrac{1}{t}$, which should be equal to the slope of $\overline{BC}$, given by $\dfrac{1}{100 - 10t}$. Therefore, $100 - 10t = t$. Hence, $100 = 11t$, so $t = \dfrac{100}{11} = 9\dfrac{1}{11}$ unit of time.

22 The point of intersection is located at $(1.5, 1)$.

23

(a) $2x - 7y = -3$

(b) $y = -2x + 2$

24 Since $\dfrac{3}{A} = \dfrac{-B}{2} = \dfrac{14}{21}$, so $A = \dfrac{9}{2}$ and $B = -2$.

1 Follow the following steps.
#1. If $a \leq b$ and $a \geq b$, then $a = b$.
#2. Hence, $a^2 - b^2 = (a-b)(a+b) = 0(a+b) = 0$.

2

(a) $2^{-3} > 3^{-3}$

(b) $\dfrac{12}{5} > \dfrac{9}{7}$

(c) Since $J > B$, then $J - 1,000,000,000 > B - 1,000,000,000$.

3 For any real number x, $x^2 \geq 0$. Let's look at three possible cases.
#1. If $x > 0$, then it is obvious.
#2. If $x = 0$, then $x^2 = 0$.
#3. If $x < 0$, then $x^2 > 0$.
Hence, $x^2 \geq 0$ for any real number x.

4 $5^{40} < 2^{100} < 3^{80}$.

5 First, $A = 1 - \dfrac{1}{54322}$ and $B = 1 - \dfrac{1}{5433}$. Hence, $A > B$.

6 $\sqrt{4\sqrt{3}} > \sqrt{2\sqrt{5}}$, since $48 > 20$.

7

$$\frac{1}{1+2+\cdots+2019} > \frac{1}{2+3+\cdots+2020}$$

because $1 + 2 + \cdots + 2019 < 2 + 3 + \cdots + 2020$.

8 $2^{85} > 2^{80} > 3^{50} > 3^{45}$.

9 $3x - 7 \geq 8 - 2x$ implies $x \geq 3$.

10

(a) $2x + 15 \leq 21 \implies 2x \leq 6 \implies x \leq 3$

(b) $11 - 2t < 19 + 6t \implies -8 < 8t \implies -1 < t$

11 $2 + x > 5 - 3x > -4$ implies $\dfrac{3}{4} < x < 3$.

12 $10 < 3\sqrt{x} - 2 < 16$ implies $4 < \sqrt{x} < 6$. Hence, $16 < x < 36$.

13 $\dfrac{1}{2} < \dfrac{x}{x+1} < \dfrac{99}{100}$ implies $1 < x < 99$.

14 Similar to #13, $1 < n < 9$. The largest n must be 8.

15

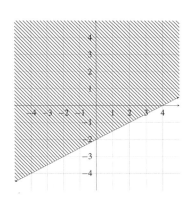

16

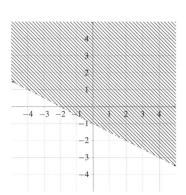

17

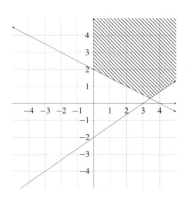

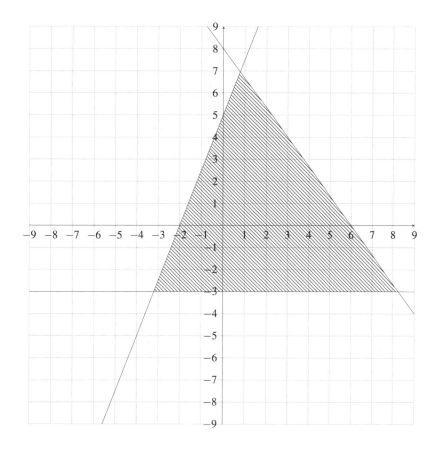

Since we want to maximize $3x+y$, put the left-most vertex of the shaded triangle, i.e., $\left(\dfrac{33}{4}, -3\right)$ into $3x+y$. Hence, the maximum value must be $\dfrac{99}{4} - 3 = \dfrac{87}{4}$.

1 Since $x^2 = 16$, try $x = 4$ and $x = -4$. Both work. They are the solutions to the equation $x^2 = 16$.

2 Since $2x^2 = 50$, try $x = 5$ and $x = -5$. Both numbers work. Hence, they are the solutions to the equation $2x^2 - 50 = 0$.

3

(a) $x = 3$ or $x = -5$ by zero-product property.

(b) $x = 4$, repeated twice, which is called the double root.

(c) $x = \dfrac{2}{3}$ or $x = -\dfrac{4}{3}$ by zero-product property.

(d) $x = 2$ or $x = -2$ by zero-product property.

4

$$10000x = \frac{4}{x}$$
$$10000x^2 = 4$$
$$(100x)^2 = 4$$
$$100x = 2$$
$$x = \frac{2}{100}$$
$$x = \frac{1}{50}$$

5

$$x^2 + 8x + 15 = 0$$
$$(x+3)(x+5) = 0$$

Hence, $x = -3$ or $x = -5$.

6

$$x^2 - 5x - 24 = 0$$
$$(x-8)(x+3) = 0$$

Hence, $x = 8$ or $x = -3$

7

(a) $x^2 - 11x + 28 = (x-4)(x-7) = 0$, so $x = 4$ or $x = 7$.

(b) $x^2 - 10x + 25 = (x-5)(x-5) = 0$, so $x = 5$ (double root).

(c) $x^2 - x - 56 = (x-8)(x+7) = 0$, so $x = 8$ or $x = -7$.

(d) $x^2 + 8x = x(x+8) = 0$, so $x = 0$ or $x = -8$.

8 $(a, b, c) = (3, 19, 6)$ by expansion. Hence, $x = -\dfrac{1}{3}$ or $x = -6$.

9

(a) $ac = 5$ and $bd = -35$

(b) $ad + bc = 18$

(c) $(5x - 7)(x + 5) = 0$, so $x = \dfrac{7}{5}$ or $x = -5$.

10

(a) $ac = 12$, $bd = -15$, and $ad + bc = 8$.

(b) $(2x + 3)(6x - 5) = 0$, so $x = -\dfrac{3}{2}$ or $x = \dfrac{5}{6}$.

11

- $(2x - 1)(2x + 2)$ has even coefficient of x.

- $(2x - 17)(2x + 3)$ has even coefficient of x.

- $(4x + 1)(x + 3)$ has odd coefficient of x.

12

(a) The coefficient of x is odd, but the product on the right side result in even coefficient of x because of $4x$ and $2x$. This is how Bob figured out the factorization in the question does not work.

(b) $12x^2 + 28x - 5 = (2x + 5)(6x - 1) = 0$, so $x = -\dfrac{5}{2}$ and $x = \dfrac{1}{6}$.

13 Follow the following steps.

\#1. Do not expand the product.

\#2. Instead, simplify the expression such that

$$(3x-2)(2x-5)+(x-7)(3x-2) = (3x-2)(2x-5+x-7)$$
$$= (3x-2)(3x-12)$$
$$= 0$$

\#3. Hence, $x = \dfrac{2}{3}$ or $x = 4$.

14 There are two possibilities.

\#1. If $5x^2+kx+2 = (5x+2)(x+1)$, then $k = 7$.

\#2. On the other hand, if $5x^2+kx+2 = (5x+1)(x+2)$, then $k = 11$.

15

(a) $x^2-7x+12 = (x-3)(x-4) = 0$, so $x = 3$ or $x = 4$.

(b) The sum of the roots is 7 and the product of roots is 12.

16 Since $-A = -2+5$ and $B = (-2)(5)$, we get $A = -3$ and $B = -10$.

17

\#1. First, $r+s = -\dfrac{2}{19}$ and $rs = \dfrac{38}{19} = 2$.

\#2. Then, $(r-1)(s-1) = rs-(r+s)+1 = 2+\dfrac{2}{19}+1 = 3+\dfrac{2}{19} = \dfrac{59}{19}$.

18

\#1. First, $r+s = 4$ and $rs = 2$.

\#2. Therefore, $n = (r+\dfrac{1}{s})(s+\dfrac{1}{r}) = rs+2+\dfrac{1}{rs} = 2+2+\dfrac{1}{2} = 4+\dfrac{1}{2} = \dfrac{9}{2}$.

1

(a) $(x+5)^2 = x^2 + 10x + 25$

(b) $(2x-3)^2 = 4x^2 - 12x + 9$

(c) $\left(\dfrac{x}{2} - 3\right)^2 = \dfrac{x^2}{4} - 3x + 9$

(d) $(5-2x)^2 = 4x^2 - 20x + 25$

2

(a) This is not a perfect square.

(b) This is a perfect square, i.e., $(z-7)^2$

(c) This is a perfect square, i.e., $\left(x - \dfrac{1}{2}\right)^2$

(d) This is a perfect square, i.e, $(-2x+7)^2$ or $(2x-7)^2$

3

Since $x = \dfrac{\sqrt{6}}{3}$, $\left(x - \dfrac{1}{x}\right)^2 = x^2 - 2 + \dfrac{1}{x^2} = \dfrac{1}{6}$

4 Using the perfect square, the answer must be 1.

5

(a) $61^2 = (60+1)^2 = (60)^2 + 2(60) + 1 = 3600 + 120 + 1 = 3721$

(b) $299^2 = (300-1)^2 = (300)^2 - 2(300) + 1 = 89401$

6

(a) $(x+y)^3 = x^3 + 3x^2y + 3xy^2 + y^3$

(b) $(x+y)^4 = x^4 + 4x^3y + 6x^2y^2 + 4xy^3 + y^4$

(c) $(x+y)^5 = x^5 + 5x^4y + 10x^3y^2 + 10x^2y^3 + 5xy^4 + y^5$

7

(a) $x^2 - 4 = (x-2)(x+2)$

(b) $2x^2 - 8 = 2(x^2 - 4) = 2(x-2)(x+2)$

(c) $25x^2 - 16y^2 = (5x - 4y)(5x + 4y)$

(d) $144x^2 - 196y^2 = (12x - 14y)(12x + 14y)$

8

(a) $6^2 - 5^2 = (6-5)(6+5 = 11$

(b) $7^2 - 6^2 = (7-6)(7+6) = 13$

(c) $11^2 - 10^2 = (11-10)(11+10) = 21$

(d) $2020^2 - 2019^2 = (2020 - 2019)(2020 + 2019) = 4039$

9

$$\sqrt{(n+3)(n+1)(n-1)(n-3) + 16} = \sqrt{(n^2 - 1)(n^2 - 9) + 16}$$
$$= \sqrt{n^4 - 10n^2 + 25}$$
$$= (n^2 - 5)$$

Since $n = 2019$ for our example, we simply compute $(2019)^2 - 5$ as the answer.

10 Since $mn + m + n = 76$, $(m+1)(n+1) = 77$. Hence, $(m,n) = (6, 10)$ or $(10, 6)$.

11 Since $bc - 7b + 3c = 70$, $(b+3)(c-7) = 49$. Therefore, $(b,c) = (-2, 56)$, $(4, 14)$, $(46, 8)$, $(-4, -42)$, $(-10, 0)$, $(52, 6)$.

12 Since $\dfrac{4}{m} + \dfrac{2}{n} = 1$, then $(m-4)(n-2) = 8$. Hence, $(m,n) = (12, 3)$, $(8, 4)$, $(6, 5)$, and $(5, 10)$.

13 Since $pq - 3p + 5q = 0$, then $(p+5)(q-3) = -15$. Hence, $(p,q) = (-4, -12)$, $(-2, -2)$, $(0, 0)$, $(6, 2)$, $(-6, 18)$, $(-8, 8)$, $(-10, 6)$, and $(-20, 4)$.

1

(a) $(-2i)^2 = (-2i)(-2i) = 4i^2 = -4$ (b) $(3i)^2 = (3i)(3i) = 9i^2 = -9$

(c) $i^{100} = (i^4)^{25} = 1^{25} = 1$ (d) $i^{-20} = \dfrac{1}{i^{20}} = \dfrac{1}{1} = 1$

2

(a) $x^2 + 25 = 0$ implies $x = \pm 5i$.   (b) $x^2 + 15 = 0$ implies $x = \pm\sqrt{15}i$.

3

(a) $i^{-124} = \dfrac{1}{i^{124}} = \dfrac{1}{1} = 1$ (b) $i^{2019} = \cdots = i^3 = -i$

4

(a) 7   (b) 6   (c) 6   (d) 0

5

(a) $(2-i)(3-2i) = 6 - 4i - 3i + 2i^2 = 4 - 7i$

(b) $(4-i)(4+i) = 4^2 - i^2 = 17$

6 $(2+5i)(2+Ai) = 4 + (2A+10)i - (5A)$ has to be real, so $2A + 10 = 0$. Hence, $A = -5$.

7

$$\frac{1}{2-i} = \frac{2+i}{(2-i)(2+i)}$$
$$= \frac{2+i}{5}$$
$$= \frac{2}{5} + \frac{1}{5}i$$

8

$$\frac{1+i}{3-i} = \frac{(1+i)(3+i)}{(3-i)(3+i)}$$
$$= \frac{2+4i}{10}$$
$$= \frac{1}{5} + \frac{2}{5}i$$

Solution for Topic 10 Questions

1 For this problem, we use that $\sqrt{x^2} = |x|$, and $|x| = a$ implies $x = a$ or $-a$, which can be abbreviated as $x = \pm a$.

(a)

(b)

$$x^2 - 16 = 0$$
$$x^2 = 16$$
$$\sqrt{x^2} = \sqrt{16}$$
$$|x| = 4$$
$$x = \pm 4$$

$$(x+2)^2 - 16 = 0$$
$$(x+2)^2 = 16$$
$$\sqrt{(x+2)^2} = \sqrt{16}$$
$$|x+2| = 4$$
$$x+2 = \pm 4$$
$$x = -2 \pm 4$$
$$x = 2 \text{ or } -6$$

(c)

(d)

$$x^2 - 12 = 0$$
$$x^2 = 12$$
$$\sqrt{x^2} = \sqrt{12}$$
$$|x| = 2\sqrt{3}$$
$$x = \pm 2\sqrt{3}$$

$$(x+2)^2 - 12 = 0$$
$$(x+2)^2 = 12$$
$$\sqrt{(x+2)^2} = \sqrt{12}$$
$$|x+2| = 2\sqrt{3}$$
$$x+2 = \pm 2\sqrt{3}$$
$$x = -2 \pm 2\sqrt{3}$$

2

$$x^2 + 2x - 7 = 0$$
$$x^2 + 2x = 7$$
$$x^2 + 2x + 1 = 8$$
$$(x+1)^2 = 8$$
$$|x+1| = 2\sqrt{2}$$
$$x+1 = \pm 2\sqrt{2}$$
$$x = -1 \pm 2\sqrt{2}$$

#1. First, let's look at the given condition $x^2 + 32x + c = (x+a)^2$.

#2. Expand the rightside :

$$x^2 + 32x + c = x^2 + 2a + a^2$$
$$32x + c = 2a + a^2$$

Here, the coefficient of x must have the same value, and $c = a^2$.

#3. Hence, $a = 16$ and $c = 16^2 = 256$.

4

(a)

$$(x+a)^2 = x^2 - 6x + c$$
$$x^2 + 2ax + a^2 = x^2 - 6x + c$$
$$2a = -6$$
$$a = -3$$
$$c = 9$$

(b)

$$(r+a)^2 = r^2 + 3r + c$$
$$r^2 + 2ar + a^2 = r^2 + 3r + c$$
$$2a = 3$$
$$a = \frac{3}{2}$$
$$c = \frac{9}{4}$$

(c)

$$(x+a)^2 = x^2 - \frac{x}{2} + c$$
$$x^2 + 2ax + a^2 = x^2 - \frac{1}{2}x + c$$
$$2a = -\frac{1}{2}$$
$$a = -\frac{1}{4}$$
$$c = \frac{1}{16}$$

5

$$x^2 + 8x = 14$$
$$x^2 + 8x + 16 = 14 + 16$$
$$(x+4)^2 = 30$$
$$x + 4 = \pm\sqrt{30}$$
$$x = -4 \pm \sqrt{30}$$

6

$$3x^2 + 12x + 1 = 0$$
$$3(x^2 + 4x) = -1$$
$$3(x^2 + 4x + 4) = -1 + 12$$
$$3(x+2)^2 = 11$$
$$(x+2)^2 = \frac{11}{3}$$
$$|x+2| = \frac{\sqrt{33}}{3}$$
$$x + 2 = \pm\frac{\sqrt{33}}{3}$$
$$x = -2 \pm \frac{\sqrt{33}}{3}$$

7

(a)

$$x^2 + 2x - 13 = 0$$
$$x^2 + 2x + 1 = 14$$
$$x + 1 = \pm\sqrt{14}$$
$$x = -1 \pm \sqrt{14}$$

(b)

$$12x^2 - 11x - 36 = 0$$
$$12(x^2 - \frac{11}{12}x) = 36$$
$$(x^2 - \frac{11}{12}x) = 3$$
$$x^2 - \frac{11}{12}x + \frac{121}{576} = 3 + \frac{121}{576}$$
$$(x - \frac{11}{24})^2 = \frac{1849}{576}$$
$$x - \frac{11}{24} = \frac{43}{24}$$
$$x = \frac{9}{4} \text{ or } -\frac{4}{3}$$

8

$$ax^2 + bx + c = 0$$
$$ax^2 + bx = -c$$
$$a(x^2 + \frac{b}{a}x) = -c$$
$$x^2 + \frac{b}{a}x = -\frac{c}{a}$$
$$x^2 + \frac{b}{a}x + \frac{b^2}{4a^2} = -\frac{c}{a} + \frac{b^2}{4a^2}$$
$$(x + \frac{b}{2a})^2 = \frac{b^2 - 4ac}{4a^2}$$
$$x + \frac{b}{2a} = \pm\frac{\sqrt{b^2 - 4ac}}{2a}$$
$$x = \frac{-b \pm \sqrt{b^2 - 4ac}}{2a}$$

9

(a) $x = 8$ or -12

(b) $x = \dfrac{7}{3}$

(c) $x = \dfrac{2}{3}$ or 4

(d) $x = -5i$ or $-i$

10 Bob's statement is false because $x = \dfrac{-1 \pm \sqrt{1 - 44}}{2}$ where the discriminant is $1 - 44 < 0$. Hence, there is no real solution to $x^2 + x + 11 = 0$.

11 Given $ax^2 - 5x + 6 = 0$, $b^2 - 4ac = (-5)^2 - 4(6)(a) = 0$, so $a = \dfrac{25}{24}$.

12 Since $x = 1 - 3i$, then $x - 1 = -3i$, so $x^2 - 2x + 1 = -9$. Therefore, $x^2 - 2x + 10 = 0$. Hence, $b = -2$ and $c = 10$.

13 Let s and t be the solutions to the equation $2x^2 - mx - 8 = 0$. Then, $s + t = \dfrac{m}{2}$ and $st = -4$. By assumption, $s - t = m - 1$. Hence, $s = \dfrac{3m}{4} - \dfrac{1}{2}$ and $t = \dfrac{1}{2} - \dfrac{m}{4}$. Therefore, $st = -4$ implies that $m = -\dfrac{10}{3}$ or $m = 6$. If $m = -\dfrac{10}{3}$, then $s - t < 0$, so $s < t$. Likewise, if $m = 6$, then $s - t > 0$, so $s > t$.

$$x^2 + bx + c = 0$$
$$(x - r)(x - s) = 0$$
$$x^2 - (r + s)x + rs = 0$$
$$r + s = -b$$
$$rs = c$$

Hence, $(r - 1)(s - 1) = 11$ implies that $rs - (r + s) + 1 = 11$, so $c - (-b) = 10$.
Hence, $b + c = 10$.

Solution for Topic 11 Questions

1

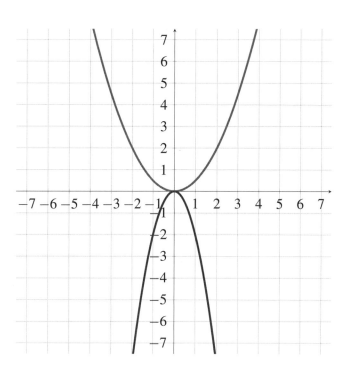

2

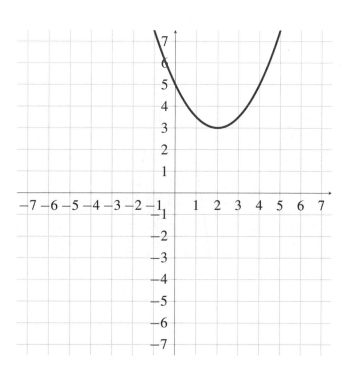

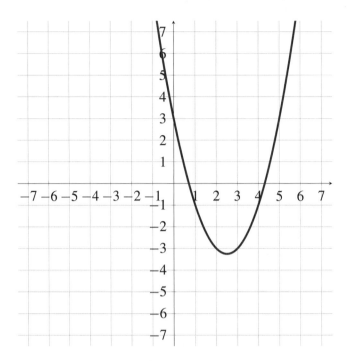

4

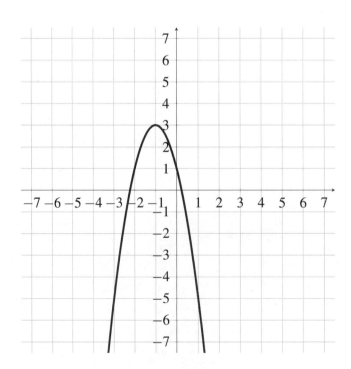

$\boxed{5}$

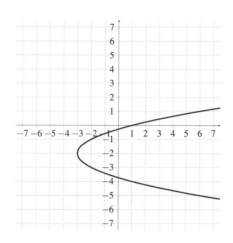

$\boxed{6}$ The circle equation is $(x-3)^2 + (y-4)^2 = 25$.

$\boxed{7}$ The center is $(2,3)$ and the radius is 6.

$\boxed{8}$ The center is $(2,4)$ and the radius is 5.

$\boxed{9}$ $(x+1)^2 + (y-\frac{3}{2})^2 = \frac{31}{12}$. Hence, the center is $(-1,\frac{3}{2})$, and the radius is $\frac{\sqrt{93}}{6}$.

$\boxed{10}$ If the circle passes through $(0,0)$, $(3,-1)$ and $(-1,-7)$, we get the following three equations by substituting the three points into the equation $(x-h)^2 + (y-k)^2 = r^2$.

$$h^2 + k^2 = r^2$$
$$(3-h)^2 + (-1-k)^2 = r^2$$
$$(-1-h)^2 + (-7-k)^2 = r^2$$

Solving the system of equations as shown in the example right above this question, we get $(h,k) = (5/11, -40/11)$ and $r = 5\sqrt{65}/11$. Hence, the circle equation must be

$$(x-5/11)^2 + (y+40/11)^2 = (5\sqrt{65}/11)^2$$

$\boxed{11}$ Substituting $x = y + 8$ into $x^2 + y^2 - 4x + 12y + 28 = 0$, then

$$(x,y) = (2+\sqrt{6}, -6+\sqrt{6}), (2-\sqrt{6}, -6-\sqrt{6})$$

$\boxed{12}$ The x-coordinate of the center is $x = 5$.

13 Given $(x+2)(x-1) \geq 0$, find the possible candidates $x = -2$ and 1.

- $x < -2 : (x+2)(x-1) > 0$
- $x = -2 : (x+2)(x-1) = 0$
- $-2 < x < 1 : (x+2)(x-1) < 0$
- $x = 1 : (x+2)(x-1) = 0$
- $x > 1 : (x+2)(x-1) > 0$

Hence, $x \leq -2$ or $1 \leq x$.

14

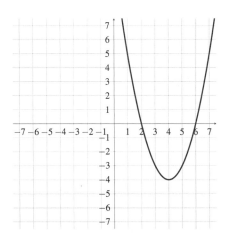

We are looking for portions of the graph that are below the x-axis. That means, the x-values range from 2 to 6. Hence, $2 < x < 6$.

15

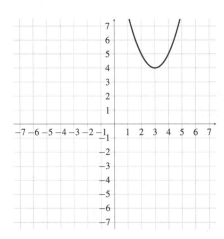

The graph of $y = x^2 - 6x + 13$ is always above the x-axis. Hence, there is no real solution satisfying $x^2 - 6x + 13 < 0$.

16

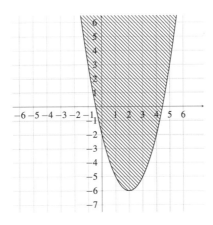

17

$$x^2 - 6x + k \geq 5$$
$$x^2 - 6x + 9 - 9 + k \geq 5$$
$$(x-3)^2 + (k-9) \geq 5$$

This implies that $k - 9 \geq 5$. Hence, $k \geq 14$.

18 Given two real numbers, x and y,

$$(x-y)^2 \geq 0$$
$$x^2 - 2xy + y^2 \geq 0$$
$$x^2 + y^2 \geq 2xy$$
$$\frac{x^2 + y^2}{2} \geq xy$$

19 Let x be a positive real number. Then, by AM-GM inequality,

$$x + \frac{1}{x} \geq 2\sqrt{x \times \frac{1}{x}} = 2$$

20 Follow the following lines of thoughts.

#1. $-x^2 + 6x - 7 = -(x-3)^2 + 2$.

#2. $(x-3)^2 \geq 0$ by trivial inequality.

#3. $-(x-3)^2 + 2 \leq 0 + 2$.

#4. We get the maximum value of 2 at $x = 3$.

21

(a) $2x^2 + 8x - 9 = 2(x+2)^2 - 17$, so the minimum value is -17 for all real x.

(b) The minimum value of y occurs at $x = 0$ because the restricted domain is in the right side of the axis of symmetry. In this region, the function is increasing so the minimum occurs at $x = 0$ with the value of -9.

22

(a)

$$-16t^2 + 48t + 45 = 0$$
$$-(4t+3)(4t-15) = 0$$
$$t = -3/4 \text{ or } 15/4$$

Since $t \geq 0$, the ball hits the ground at $t = 15/4$ seconds.

(b)

$$-16t^2 + 48t + 45 = 0$$
$$-16(t^2 - 3t) + 45 = 0$$
$$-16(t^2 - 3t + 9/4 - 9/4) + 45 = 0$$
$$-16(t - 3/2)^2 + 81 = 0$$

Hence, the maximum height is 81 feet at $t = 3/2$ seconds.

23 Follow the following instructions.

#1. Let x be the length and y be the width of the barn. Assume that there is only one side for length and two sides for width.

#2. Set up a proper equation : $2x + y = 20$.

#3. Maximize the area : xy.

#4. Find the maximum value of
$x(20 - 2x) = 20x - 2x^2 = -2(x^2 - 10x + 25 - 25) = -2(x-5)^2 + 50$.

#5. Conclude that the maximum area is 50 square meters.

1

(a) $f(1) = 4(1) - 2 = 2$, $f(2) = 4(2) - 2 = 6$, $f(3x) = 4(3x) - 2 = 12x - 2$

(b) $f(0) = -2$

(c) $f(2k - 3) = 4(2k - 3) - 2 = 8k - 14$

2

(a) $f(1) = f(-1) = 4$, and $f(-2) = 7$

(b) $f(x) = 4$ implies $x = \pm 1$. There is no real x-value for $f(x) = 2$.

(c) The range is $[3, \inf)$, and the domain is $\mathbb{R}$.

3

(a) $f(1) = -\dfrac{1}{5}$, $f(0) = -\dfrac{2}{3}$, and $f(-\dfrac{1}{2}) = -\dfrac{5}{4}$

(b) $x = -\dfrac{3}{2}$. Hence, the domain is the set of all real numbers except $x = -\dfrac{3}{2}$.

(c) There is no value of x such that $f(x) = \dfrac{1}{2}$ because $x = -\dfrac{-2 - 3y}{2y - 1}$ where $2y \neq 1$. Hence, the range is the set of all real numbers except $y = \dfrac{1}{2}$.

4

(a) $x = \dfrac{1}{3}$ is the solution for $2 - 3x = 1$, which is not in the domain of $h(x)$. Hence, 1 is not in the range of $h(x)$.

(b) The range of $h(x)$ is $[-25, -4]$.

5 $g(1, y, 2) = 2(1) - 3(2) + 4y = 11$, so $y = \dfrac{15}{4}$.

6 $x^2 + 2x + y^2 - 4y + 5 \geq 0$ for all x and y. Hence, $f(x, y) \geq 0$.

7

(a) $f(1) + g(1) = 1$

(b) $f(2) - g(2) = 18$

(c) $f(4) \cdot g(4) = -275$

(d) $\dfrac{f(2)}{g(2)} = -\dfrac{7}{11}$

8

(a) $[0, \infty)$

(b) $(-\infty, -2] \cup [3, \infty)$

(c) $[3, \infty)$

(d) $(3, \infty)$

9

(a) $[0, 2]$

(b) $[0, 2]$

(c) $(0, 2]$

10

(a) $f(g(1)) = f(1) = -2$

(b) $g(f(1)) = g(-2) = 10$

(c) $f(g(ax + b)) = (4 - 3ax - 3b) - 3 = 1 - 3ax - 3b$

11 $f(f(x)) = 2(2x + 10) + 10 = 4x + 20 + 10 = 4x + 30 = x$, so $3x + 30 = 0$. Hence, $x = -10$.

12 Since $f(x) = x - 3$, $f(f(x)) = f(x - 3) = x - 6 = 4$. Hence, $x = 10$.

13 $f^{-1}(x) = \dfrac{x + 9}{2}$ using $f(f^{-1}(x)) = x$.

14 Let $f(4) = x$. Then, $4 = f^{-1}(x) = \dfrac{2x - 3}{x + 5}$. Hence, $4(x + 5) = 2x - 3$.

Therefore, $x = -\dfrac{23}{2}$.

15

(a) $f(f^{-1}(x)) = x$ implies that $f(x) = \pm\sqrt{x}$, which means that this is not a function.

(b) If $g(x) = x^2$ for $x \geq 0$. Then, the $g^{-1}(x) = \sqrt{x}$, which is a function, indeed.

16

$$f(2) = 5$$
$$2 = f^{-1}(5)$$

Solution for Topic 13 Questions

1

(a)

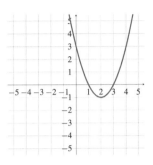

(b) The x-intercepts are 1 or 3.

(c) The y-intercept is 3.

2

(a) $f(3) = 6$ and $f(f(3)) = f(6) = 2$

(b) The domain is $[-6, 7]$, meaning that $-6 \leq x \leq 7$.

(c) The range is $[-3, 6]$, meaning that $-3 \leq y \leq 6$.

3

(a) The graph represents a function $y = \sqrt{x}$. This is a function known as a square root function.

(b) The graph does not represent a function. In fact, this is the graph of $x = y^2$, which is a parabola. Draw a vertical line at $x = 1$, for instance. The line has two points of intersection, as shown in the graph below. Hence, this does not pass VLT(vertical line test), which means the graph cannot be that of a function.

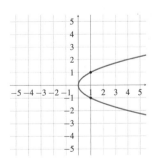

(a) $f(x) = 1$

(b) $f(x) = x$

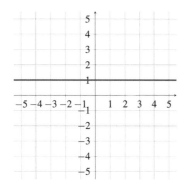

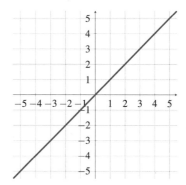

(c) $f(x) = x^2 + 1$

(d) $f(x) = \sqrt{x}$

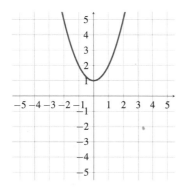

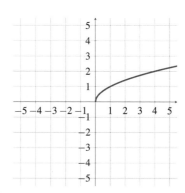

(e) $f(x) = \dfrac{1}{x}$

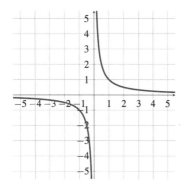

5

#1. The graph belongs to that of a function.

#2. The domain is $\{-5, -4, -3, -2, -1, 0, 1, 2, 3, 4, 5\}$.

#3. Similarly, the range is $\{-3, -1, 0, 1, 2, 3\}$.

6

(a) $y = f(2x)$

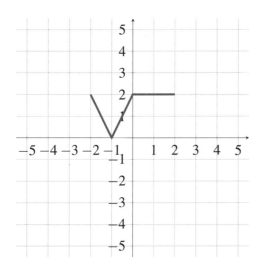

(b) $y = 2f(x)$

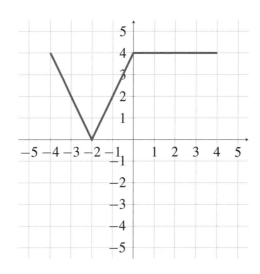

7

(a) $y = f(\frac{1}{2}x)$

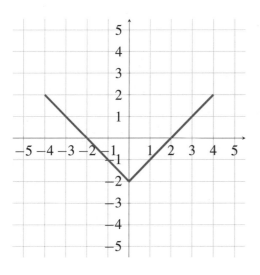

(b) $y = \frac{1}{2}f(x)$

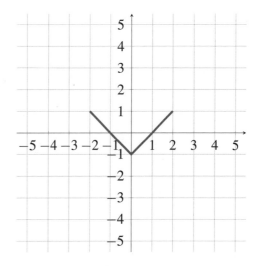

8

(a) $y = f(-x)$

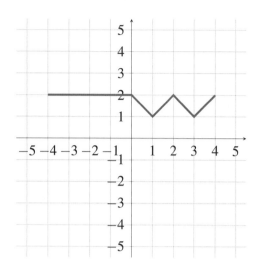

(b) $y = -f(x)$

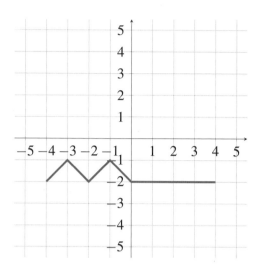

9

(a) $y = f(x) + 3$

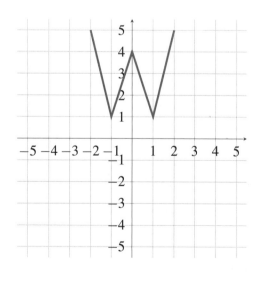

(b) $y = f(x - 2)$

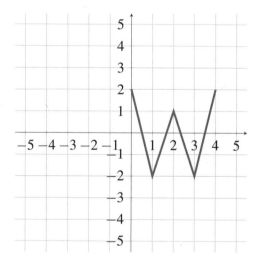

The graph of $y = 2f(x-2) - 1$ is found by

- vertically stretching the graph of $y = f(x)$ by 2.

- horizontally shifting the graph of $y = 2f(x)$ right by 2.

- vertically shifting the graph of $y = 2f(x-2)$ down by 1.

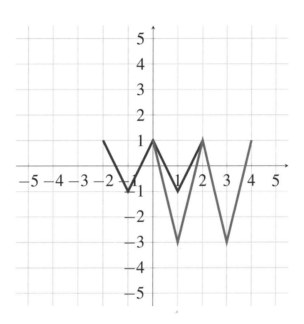

(a) $(2,4) \to (1,4) \to (1,6)$. Check that $6 = f(2(1)) + 4$, so $(1,6)$ is on the graph of $y = f(2x) + 2$.

(b) $(2,4) \to (-2,4) \to (-2,5)$. Check that $5 = f(-(-2)) + 1$, so $(-2,5)$ is on the graph of $y = f(-x) + 1$.

(c) $(2,4) \to (4,4) \to (4,3)$. Check that $3 = f\left(\dfrac{4}{2}\right) - 1$, so $(4,3)$ is on the graph of $y = f\left(\dfrac{x}{2}\right) - 1$.

(d) $(2,4) \to (1,8) \to (3,8) \to (3,12)$. Check that $12 = 2f(2(3) - 4) + 4$, so $(3,12)$ is on the graph of $y = 2f(2x-4) + 4$.

(a) Since $f(f^{-1}(x)) = x$,

$$f(f^{-1}(x)) = 2f^{-1}(x) + 1$$
$$x = 2f^{-1}(x) + 1$$
$$x - 1 = 2f^{-1}(x)$$
$$\frac{x-1}{2} = f^{-1}(x)$$

Hence, $f^{-1}(x) = \dfrac{x}{2} - \dfrac{1}{2}$.

(b)

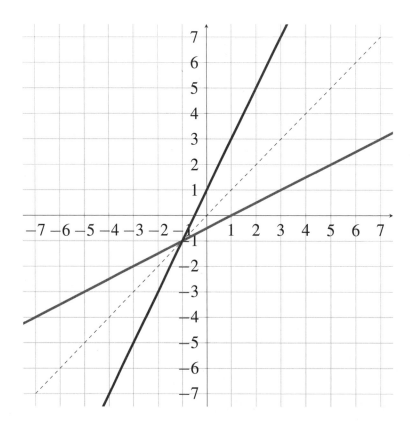

The blue line is the graph of $y = f(x) = 2x + 1$ and the red line is the graph of $y = f^{-1}(x) = \dfrac{x}{2} - \dfrac{1}{2}$. The two lines are symmetric about the line $y = x$.

13

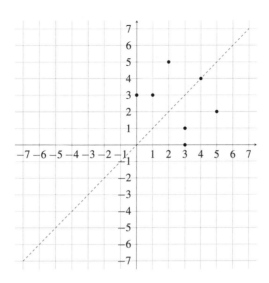

14

(a) The graph does not pass HLT, so it does not have an inverse function.

(b) The graph passes HLT, so it has an inverse function. In fact, this is the graph of $y = x^3$.

15

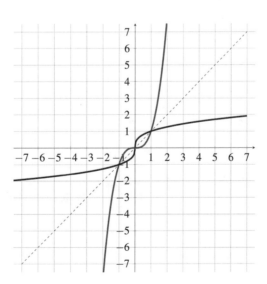

Solution for Topic 14 Questions

1

(a) $f(x) + g(x) = -2x^3 + 2x^2 - 3x$

(b) $f(x) - g(x) = 4x^3 - 2x^2 - 5x + 14$

2

$$(3x^2 - 2x + 3)(x^3 - 2x^2 + x - 7) = 3x^5 + (-6 - 2)x^4 + (3 + 4 + 3)x^3$$
$$+ (-21 - 2 - 6)x^2 + (14 + 3)x + 3(-7)$$
$$= 3x^5 - 8x^4 + 10x^3 - 29x^2 + 17x - 21$$

3

#1. Look at the coefficient of x^4. Since $a + b = 13$, then $(a + b)^2 = 169$.
#2. Using the coefficient of x^3, we get $ab = 40$. Hence,
$a^2 + b^2 = (a + b)^2 - 2ab = 169 - 80 = 89$.

4 $ax^3 + (-6)x^3 = (a - 6)x^3 = -11x^3$, so $a - 6 = -11$. Hence, $a = -5$.

5

(a) $\dfrac{x - 2}{3x^2 - 12} = \dfrac{x - 2}{3(x^2 - 4)} = \dfrac{x - 2}{3(x - 2)(x + 2)} = \dfrac{1}{3(x + 2)}$ where $x \neq 2$.

(b) $\dfrac{x - 1}{x^2 - 5x + 4} = \dfrac{x - 1}{(x - 4)(x - 1)} = \dfrac{1}{x - 4}$ where $x \neq 1$.

(c) $\dfrac{x^2 - 5x + 14}{x^2 + 4x + 4} = \dfrac{(x - 7)(x + 2)}{(x + 2)^2} = \dfrac{x - 7}{x + 2}$

(d) $\dfrac{x^3 - x^2 - 42x}{2x^2 - 20x + 42} = \dfrac{x(x^2 - x - 42)}{2(x^2 - 10x + 21)} = \dfrac{x + 6}{x - 3}$ where $x \neq 0, 7$

(e) $\dfrac{9x^2 - 81x}{x^3 - 8x^2 - 9x} = \dfrac{9}{x + 1}$ where $x \neq 0, 9$

(f) $\dfrac{x^2 - 2x - 8}{2x^3 - 24x^2 + 64x} = \dfrac{(x - 4)(x + 2)}{2x(x - 4)(x - 8)} = \dfrac{x + 2}{2x(x - 8)}$ where $x \neq 4$

▶ Solution for Topic 14 Questions

(a) $x = \dfrac{1}{3}$

(b) $x = \dfrac{1}{2}$

(c) $x = \pm 3$

(d) No excluded value of x

(e) $x = \pm \dfrac{7}{5}$

(f) No excluded value of x

7

(a) $x^3 - 2x^2 - 14x - 5 = (x+3)(x^2 - 5x + 1) + (-8)$

(b) $x^3 - 4x^2 - 30x + 18 = (x-2)(x^2 - 2x - 34) + (-50)$

(c) $x^3 + 5x^2 - 16x + 7 = (x-4)(x^2 + 9x + 20) + 87$

(d) $12x^3 - 4x^2 + 5x + 2 = (2x-1)(6x^2 + x + 3) + 5$

8

(a) $\dfrac{2}{3x+3} + \dfrac{1}{x-4} = \dfrac{5x-8}{3x^2 - 9x - 12}$

(b) $\dfrac{5}{x-1} + \dfrac{4}{2-x} = \dfrac{5}{x-1} - \dfrac{4}{x-2} = \dfrac{x-6}{x^2 - 3x + 2}$

(c) $\dfrac{4x}{x-3} + \dfrac{3}{x+5} = \dfrac{4x^2 + 23x - 9}{x^2 + 2x - 15}$

(d) $\dfrac{5}{2-x} - \dfrac{12x}{x^2 - 4} = \dfrac{-5}{x-2} - \dfrac{12x}{x^2 - 4} = \dfrac{-17x - 10}{x^2 - 4}$

9

(a)

$$\dfrac{6}{x^2 - 4x + 4} = \dfrac{1}{x^2 - 4x + 4} - \dfrac{1}{x-2}$$

$$\dfrac{6}{(x-2)^2} = \dfrac{1}{(x-2)^2} - \dfrac{1}{x-2}$$

$$\dfrac{5}{(x-2)^2} = \dfrac{-(x-2)}{(x-2)^2}$$

$$5 = -x + 2$$

$$3 = -x$$

$$-3 = x$$

(b)

$$\dfrac{3}{x+3} + \dfrac{3x}{x+3} = 1$$

$$\dfrac{3 + 3x}{x+3} = 1$$

$$3x + 3 = x + 3$$

$$2x = 0$$

$$x = 0$$

(c)

$$\frac{x+3}{x^2-3x} = \frac{1}{x-3} - \frac{x+5}{x^2-3x}$$

$$\frac{x+3}{x^2-3x} + \frac{x+5}{x^2-3x} = \frac{1}{x-3}$$

$$\frac{2x+8}{x^2-3x} = \frac{1}{x-3}$$

$$\frac{2x+8}{x^2-3x} = \frac{x}{x^2-3x}$$

$$2x+8 = x$$

$$x = -8$$

(d)

$$\frac{4}{x} = \frac{1}{x-4} - \frac{x+2}{x^2-4x}$$

$$4(x-4) = x - (x+2)$$

$$4x - 16 = -2$$

$$4x = 14$$

$$x = \frac{7}{2}$$

$\boxed{10}$

(a)

$$\frac{x+2}{x} + \frac{4x+2}{x^2-3x} = 1 - \frac{1}{x}$$

$$\frac{4x+2}{x^2-3x} = \frac{-3(x-3)}{x(x-3)}$$

$$4x+2 = -3(x-3)$$

$$4x+2 = -3x+9$$

$$7x = 7$$

$$x = 1$$

(b)

$$\frac{1}{x} - 5 = \frac{1}{x^2+x}$$

$$\frac{1}{x} - \frac{5x}{x} = \frac{1}{x^2+x}$$

$$\frac{1-5x}{x} = \frac{1}{x^2+x}$$

$$\frac{(x+1)(1-5x)}{x(x+1)} = \frac{1}{x^2+x}$$

$$(x+1)(1-5x) = 1$$

$$-x(5x+4) = 0$$

$$x = 0, -\frac{4}{5}$$

$$x = -\frac{4}{5}$$

(c)

$$\frac{1}{2} + \frac{1}{2x} = \frac{x^2 - 7x + 10}{x - 1}$$

$$\frac{x + 1}{2x} = \frac{x^2 - 7x + 10}{x - 1}$$

$$(x - 1)(x + 1) = 2x(x - 2)(x - 5)$$

$$x^2 - 1 = 2x(x^2 - 7x + 10)$$

$$0 = 2x^3 - 15x^2 + 20x + 1$$

Hence, a root to the polynomial $2x^3 - 15x^2 + 20x + 1 = 0$ can solve the fraction. The method of finding the solutions to the cubic polynomial(of degree 3) will be further investigated and scrutinized in Algebra 2.

1 Given $\{1,2,3,3,4,5,6\}$, the mean is equal to $\dfrac{24}{7}$, the median is 3, and the mode is 3.

2 Given $\{1,2,3,4,5\}$, the standard deviation is equal to

$$\sqrt{\frac{(1-3)^2+(2-3)^2+(3-3)^2+(4-3)^2+(5-3)^2}{5}} = \sqrt{2}$$

3 Given $\{2,3,3,4,5,6\}$, the range is $6-2=4$ and the interquartile range is $5-3=2$.

4 Given $\{2019,2020,2021,2022,2022,2023\}$, the lower quartile is 2020 and the upper quartile is 2022. Hence, the interquartile range is 2, the difference between the lower quartile and the upper quartile.

5 If the dots stay close to the least square regression line with positive slope, we say the correlation is positively strong. Otherwise, we say the correlation is positively weak. If the dots, on the other hand, stay close to the least square regression line with negative slope, the correlation is negatively strong. Otherwise, it is negatively weak. In our data plots, the strongest correlation out of three plots is B, and the weakest correlation is A.

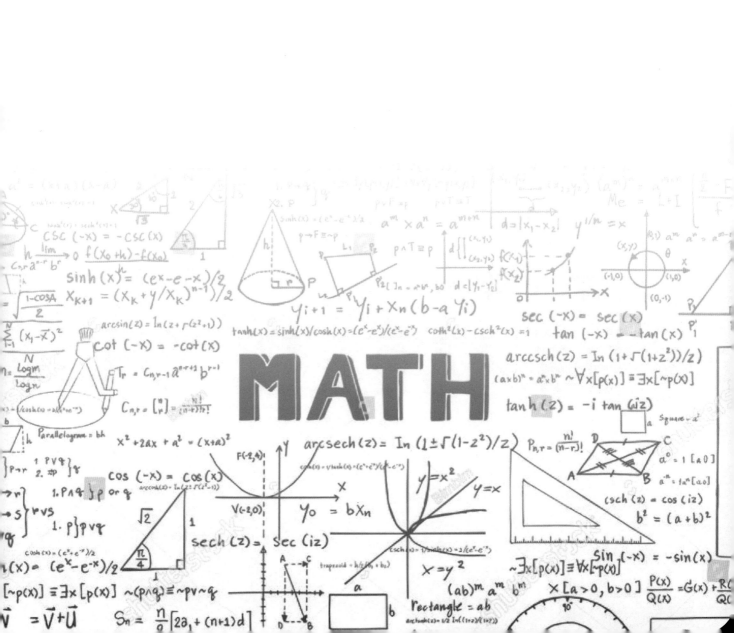

The Essential Guide to **Algebra 1**

초판 1쇄 발행　2020년 6월 10일
개정2판1쇄 발행　2023년 6월 16일

저자　유하림
발행인　최영민
발행처　헤르몬하우스
주소　경기도 파주시 신촌로 16
전화　031 – 8071 – 0088
팩스　031 – 942 – 8688
전자우편　hermonh@naver.com
출판등록　2015년 3월 27일
등록번호　제406 – 2015 – 31호
정가　20,000원

ISBN　979–11–92520–48–3　(53410)